Principles of Statistical Inference

In this important book, D. R. Cox develops the key concepts of the theory of statistical inference, in particular describing and comparing the main ideas and controversies over foundational issues that have rumbled on for more than 200 years. Continuing a 60-year career of contribution to statistical thought, Professor Cox is ideally placed to give the comprehensive, balanced account of the field that is now needed.

The careful comparison of frequentist and Bayesian approaches to inference allows readers to form their own opinion of the advantages and disadvantages. Two appendices give a brief historical overview and the author's more personal assessment of the merits of different ideas.

The content ranges from the traditional to the contemporary. While specific applications are not treated, the book is strongly motivated by applications across the sciences and associated technologies. The underlying mathematics is kept as elementary as feasible, though some previous knowledge of statistics is assumed. This book is for every serious user or student of statistics – in particular, for anyone wanting to understand the uncertainty inherent in conclusions from statistical analyses.

Principles of Statistical Inference

D.R. COX

Nuffield College, Oxford

CAMBRIDGE
UNIVERSITY PRESS

University Printing House, Cambridge CB2 8BS, United Kingdom

Cambridge University Press is part of the University of Cambridge.

It furthers the University's mission by disseminating knowledge in the pursuit of education, learning and research at the highest international levels of excellence.

www.cambridge.org
Information on this title: www.cambridge.org/9780521685672

© D. R. Cox 2006

This publication is in copyright. Subject to statutory exception and to the provisions of relevant collective licensing agreements, no reproduction of any part may take place without the written permission of Cambridge University Press.

First published 2006

A catalogue record for this publication is available from the British Library

ISBN 978-0-521-86673-6 Hardback
ISBN 978-0-521-68567-2 Paperback

Cambridge University Press has no responsibility for the persistence or accuracy of URLs for external or third-party internet websites referred to in this publication, and does not guarantee that any content on such websites is, or will remain, accurate or appropriate.

Contents

List of examples		ix
Preface		xiii
1	**Preliminaries**	1
	Summary	1
	1.1 Starting point	1
	1.2 Role of formal theory of inference	3
	1.3 Some simple models	3
	1.4 Formulation of objectives	7
	1.5 Two broad approaches to statistical inference	7
	1.6 Some further discussion	10
	1.7 Parameters	13
	Notes 1	14
2	**Some concepts and simple applications**	17
	Summary	17
	2.1 Likelihood	17
	2.2 Sufficiency	18
	2.3 Exponential family	20
	2.4 Choice of priors for exponential family problems	23
	2.5 Simple frequentist discussion	24
	2.6 Pivots	25
	Notes 2	27
3	**Significance tests**	30
	Summary	30
	3.1 General remarks	30
	3.2 Simple significance test	31
	3.3 One- and two-sided tests	35

	3.4	Relation with acceptance and rejection	36
	3.5	Formulation of alternatives and test statistics	36
	3.6	Relation with interval estimation	40
	3.7	Interpretation of significance tests	41
	3.8	Bayesian testing	42
		Notes 3	43
4	**More complicated situations**		**45**
		Summary	45
	4.1	General remarks	45
	4.2	General Bayesian formulation	45
	4.3	Frequentist analysis	47
	4.4	Some more general frequentist developments	50
	4.5	Some further Bayesian examples	59
		Notes 4	62
5	**Interpretations of uncertainty**		**64**
		Summary	64
	5.1	General remarks	64
	5.2	Broad roles of probability	65
	5.3	Frequentist interpretation of upper limits	66
	5.4	Neyman–Pearson operational criteria	68
	5.5	Some general aspects of the frequentist approach	68
	5.6	Yet more on the frequentist approach	69
	5.7	Personalistic probability	71
	5.8	Impersonal degree of belief	73
	5.9	Reference priors	76
	5.10	Temporal coherency	78
	5.11	Degree of belief and frequency	79
	5.12	Statistical implementation of Bayesian analysis	79
	5.13	Model uncertainty	84
	5.14	Consistency of data and prior	85
	5.15	Relevance of frequentist assessment	85
	5.16	Sequential stopping	88
	5.17	A simple classification problem	91
		Notes 5	93
6	**Asymptotic theory**		**96**
		Summary	96
	6.1	General remarks	96
	6.2	Scalar parameter	97

6.3	Multidimensional parameter	107
6.4	Nuisance parameters	109
6.5	Tests and model reduction	114
6.6	Comparative discussion	117
6.7	Profile likelihood as an information summarizer	119
6.8	Constrained estimation	120
6.9	Semi-asymptotic arguments	124
6.10	Numerical-analytic aspects	125
6.11	Higher-order asymptotics	128
Notes 6		130

7 Further aspects of maximum likelihood — 133

Summary — 133
- 7.1 Multimodal likelihoods — 133
- 7.2 Irregular form — 135
- 7.3 Singular information matrix — 139
- 7.4 Failure of model — 141
- 7.5 Unusual parameter space — 142
- 7.6 Modified likelihoods — 144

Notes 7 — 159

8 Additional objectives — 161

Summary — 161
- 8.1 Prediction — 161
- 8.2 Decision analysis — 162
- 8.3 Point estimation — 163
- 8.4 Non-likelihood-based methods — 169

Notes 8 — 175

9 Randomization-based analysis — 178

Summary — 178
- 9.1 General remarks — 178
- 9.2 Sampling a finite population — 179
- 9.3 Design of experiments — 184

Notes 9 — 192

Appendix A: A brief history — 194

Appendix B: A personal view — 197

References — 201

Author index — 209

Subject index — 213

List of examples

Example 1.1	The normal mean	3
Example 1.2	Linear regression	4
Example 1.3	Linear regression in semiparametric form	4
Example 1.4	Linear model	4
Example 1.5	Normal theory nonlinear regression	4
Example 1.6	Exponential distribution	5
Example 1.7	Comparison of binomial probabilities	5
Example 1.8	Location and related problems	5
Example 1.9	A component of variance model	11
Example 1.10	Markov models	12
Example 2.1	Exponential distribution (ctd)	19
Example 2.2	Linear model (ctd)	19
Example 2.3	Uniform distribution	20
Example 2.4	Binary fission	20
Example 2.5	Binomial distribution	21
Example 2.6	Fisher's hyperbola	22
Example 2.7	Binary fission (ctd)	23
Example 2.8	Binomial distribution (ctd)	23
Example 2.9	Mean of a multivariate normal distribution	27
Example 3.1	Test of a Poisson mean	32
Example 3.2	Adequacy of Poisson model	33
Example 3.3	More on the Poisson distribution	34
Example 3.4	Test of symmetry	38
Example 3.5	Nonparametric two-sample test	39
Example 3.6	Ratio of normal means	40
Example 3.7	Poisson-distributed signal with additive noise	41

Example 4.1	Uniform distribution of known range	47
Example 4.2	Two measuring instruments	48
Example 4.3	Linear model	49
Example 4.4	Two-by-two contingency table	51
Example 4.5	Mantel–Haenszel procedure	54
Example 4.6	Simple regression for binary data	55
Example 4.7	Normal mean, variance unknown	56
Example 4.8	Comparison of gamma distributions	56
Example 4.9	Unacceptable conditioning	56
Example 4.10	Location model	57
Example 4.11	Normal mean, variance unknown (ctd)	59
Example 4.12	Normal variance	59
Example 4.13	Normal mean, variance unknown (ctd)	60
Example 4.14	Components of variance	61
Example 5.1	Exchange paradox	67
Example 5.2	Two measuring instruments (ctd)	68
Example 5.3	Rainy days in Gothenburg	70
Example 5.4	The normal mean (ctd)	71
Example 5.5	The noncentral chi-squared distribution	74
Example 5.6	A set of binomial probabilities	74
Example 5.7	Exponential regression	75
Example 5.8	Components of variance (ctd)	80
Example 5.9	Bias assessment	82
Example 5.10	Selective reporting	86
Example 5.11	Precision-based choice of sample size	89
Example 5.12	Sampling the Poisson process	90
Example 5.13	Multivariate normal distributions	92
Example 6.1	Location model (ctd)	98
Example 6.2	Exponential family	98
Example 6.3	Transformation to near location form	99
Example 6.4	Mixed parameterization of the exponential family	112
Example 6.5	Proportional hazards Weibull model	113
Example 6.6	A right-censored normal distribution	118
Example 6.7	Random walk with an absorbing barrier	119
Example 6.8	Curved exponential family model	121
Example 6.9	Covariance selection model	123
Example 6.10	Poisson-distributed signal with estimated background	124
Example 7.1	An unbounded likelihood	134
Example 7.2	Uniform distribution	135
Example 7.3	Densities with power-law contact	136
Example 7.4	Model of hidden periodicity	138

Example 7.5	A special nonlinear regression	139
Example 7.6	Informative nonresponse	140
Example 7.7	Integer normal mean	143
Example 7.8	Mixture of two normal distributions	144
Example 7.9	Normal-theory linear model with many parameters	145
Example 7.10	A non-normal illustration	146
Example 7.11	Parametric model for right-censored failure data	149
Example 7.12	A fairly general stochastic process	151
Example 7.13	Semiparametric model for censored failure data	151
Example 7.14	Lag one correlation of a stationary Gaussian time series	153
Example 7.15	A long binary sequence	153
Example 7.16	Case-control study	154
Example 8.1	A new observation from a normal distribution	162
Example 8.2	Exponential family	165
Example 8.3	Correlation between different estimates	165
Example 8.4	The sign test	166
Example 8.5	Unbiased estimate of standard deviation	167
Example 8.6	Summarization of binary risk comparisons	171
Example 8.7	Brownian motion	174
Example 9.1	Two-by-two contingency table	190

Preface

Most statistical work is concerned directly with the provision and implementation of methods for study design and for the analysis and interpretation of data. The theory of statistics deals in principle with the general concepts underlying all aspects of such work and from this perspective the formal theory of statistical inference is but a part of that full theory. Indeed, from the viewpoint of individual applications, it may seem rather a small part. Concern is likely to be more concentrated on whether models have been reasonably formulated to address the most fruitful questions, on whether the data are subject to unappreciated errors or contamination and, especially, on the subject-matter interpretation of the analysis and its relation with other knowledge of the field.

Yet the formal theory is important for a number of reasons. Without some systematic structure statistical methods for the analysis of data become a collection of tricks that are hard to assimilate and interrelate to one another, or for that matter to teach. The development of new methods appropriate for new problems would become entirely a matter of ad hoc ingenuity. Of course such ingenuity is not to be undervalued and indeed one role of theory is to assimilate, generalize and perhaps modify and improve the fruits of such ingenuity.

Much of the theory is concerned with indicating the uncertainty involved in the conclusions of statistical analyses, and with assessing the relative merits of different methods of analysis, and it is important even at a very applied level to have some understanding of the strengths and limitations of such discussions. This is connected with somewhat more philosophical issues connected with the nature of probability. A final reason, and a very good one, for study of the theory is that it is interesting.

The object of the present book is to set out as compactly as possible the key ideas of the subject, in particular aiming to describe and compare the main ideas and controversies over more foundational issues that have rumbled on at varying levels of intensity for more than 200 years. I have tried to describe the various approaches in a dispassionate way but have added an appendix with a more personal assessment of the merits of different ideas.

Some previous knowledge of statistics is assumed and preferably some understanding of the role of statistical methods in applications; the latter understanding is important because many of the considerations involved are essentially conceptual rather than mathematical and relevant experience is necessary to appreciate what is involved.

The mathematical level has been kept as elementary as is feasible and is mostly that, for example, of a university undergraduate education in mathematics or, for example, physics or engineering or one of the more quantitative biological sciences. Further, as I think is appropriate for an introductory discussion of an essentially applied field, the mathematical style used here eschews specification of regularity conditions and theorem–proof style developments. Readers primarily interested in the qualitative concepts rather than their development should not spend too long on the more mathematical parts of the book.

The discussion is implicitly strongly motivated by the demands of applications, and indeed it can be claimed that virtually everything in the book has fruitful application somewhere across the many fields of study to which statistical ideas are applied. Nevertheless I have not included specific illustrations. This is partly to keep the book reasonably short, but, more importantly, to focus the discussion on general concepts without the distracting detail of specific applications, details which, however, are likely to be crucial for any kind of realism.

The subject has an enormous literature and to avoid overburdening the reader I have given, by notes at the end of each chapter, only a limited number of key references based on an admittedly selective judgement. Some of the references are intended to give an introduction to recent work whereas others point towards the history of a theme; sometimes early papers remain a useful introduction to a topic, especially to those that have become suffocated with detail. A brief historical perspective is given as an appendix.

The book is a much expanded version of lectures given to doctoral students of the Institute of Mathematics, Chalmers/Gothenburg University, and I am very grateful to Peter Jagers and Nanny Wermuth for their invitation and encouragement. It is a pleasure to thank Ruth Keogh, Nancy Reid and Rolf Sundberg for their very thoughtful detailed and constructive comments and advice on a preliminary version. It is a pleasure to thank also Anthony Edwards and Deborah Mayo for advice on more specific points. I am solely responsible for errors of fact and judgement that remain.

The book is in broadly three parts. The first three chapters are largely introductory, setting out the formulation of problems, outlining in a simple case the nature of frequentist and Bayesian analyses, and describing some special models of theoretical and practical importance. The discussion continues with the key ideas of likelihood, sufficiency and exponential families.

Chapter 4 develops some slightly more complicated applications. The long Chapter 5 is more conceptual, dealing, in particular, with the various meanings of probability as it is used in discussions of statistical inference. Most of the key concepts are in these chapters; the remaining chapters, especially Chapters 7 and 8, are more specialized.

Especially in the frequentist approach, many problems of realistic complexity require approximate methods based on asymptotic theory for their resolution and Chapter 6 sets out the main ideas. Chapters 7 and 8 discuss various complications and developments that are needed from time to time in applications. Chapter 9 deals with something almost completely different, the possibility of inference based not on a probability model for the data but rather on randomization used in the design of the experiment or sampling procedure.

I have written and talked about these issues for more years than it is comfortable to recall and am grateful to all with whom I have discussed the topics, especially, perhaps, to those with whom I disagree. I am grateful particularly to David Hinkley with whom I wrote an account of the subject 30 years ago. The emphasis in the present book is less on detail and more on concepts but the eclectic position of the earlier book has been kept.

I appreciate greatly the care devoted to this book by Diana Gillooly, Commissioning Editor, and Emma Pearce, Production Editor, Cambridge University Press.

1
Preliminaries

Summary. Key ideas about probability models and the objectives of statistical analysis are introduced. The differences between frequentist and Bayesian analyses are illustrated in a very special case. Some slightly more complicated models are introduced as reference points for the following discussion.

1.1 Starting point

We typically start with a subject-matter question. Data are or become available to address this question. After preliminary screening, checks of data quality and simple tabulations and graphs, more formal analysis starts with a provisional model. The data are typically split in two parts ($y : z$), where y is regarded as the observed value of a vector random variable Y and z is treated as fixed. Sometimes the components of y are direct measurements of relevant properties on study individuals and sometimes they are themselves the outcome of some preliminary analysis, such as means, measures of variability, regression coefficients and so on. The set of variables z typically specifies aspects of the system under study that are best treated as purely explanatory and whose observed values are not usefully represented by random variables. That is, we are interested solely in the distribution of outcome or response variables conditionally on the variables z; a particular example is where z represents treatments in a randomized experiment.

We use throughout the notation that observable random variables are represented by capital letters and observations by the corresponding lower case letters.

A model, or strictly a family of models, specifies the density of Y to be

$$f_Y(y : z; \theta), \qquad (1.1)$$

where $\theta \subset \Omega_\theta$ is unknown. The distribution may depend also on design features of the study that generated the data. We typically simplify the notation to $f_Y(y; \theta)$, although the explanatory variables z are frequently essential in specific applications.

To choose the model appropriately is crucial to fruitful application.

We follow the very convenient, although deplorable, practice of using the term *density* both for continuous random variables and for the probability function of discrete random variables. The deplorability comes from the functions being dimensionally different, probabilities per unit of measurement in continuous problems and pure numbers in discrete problems. In line with this convention in what follows integrals are to be interpreted as sums where necessary. Thus we write

$$E(Y) = E(Y; \theta) = \int y f_Y(y; \theta) dy \qquad (1.2)$$

for the expectation of Y, showing the dependence on θ only when relevant. The integral is interpreted as a sum over the points of support in a purely discrete case. Next, for each aspect of the research question we partition θ as (ψ, λ), where ψ is called the *parameter of interest* and λ is included to complete the specification and commonly called a *nuisance parameter*. Usually, but not necessarily, ψ and λ are *variation independent* in that Ω_θ is the Cartesian product $\Omega_\psi \times \Omega_\lambda$. That is, any value of ψ may occur in connection with any value of λ. The choice of ψ is a subject-matter question. In many applications it is best to arrange that ψ is a scalar parameter, i.e., to break the research question of interest into simple components corresponding to strongly focused and incisive research questions, but this is not necessary for the theoretical discussion.

It is often helpful to distinguish between the primary features of a model and the secondary features. If the former are changed the research questions of interest have either been changed or at least formulated in an importantly different way, whereas if the secondary features are changed the research questions are essentially unaltered. This does not mean that the secondary features are unimportant but rather that their influence is typically on the method of estimation to be used and on the assessment of precision, whereas misformulation of the primary features leads to the wrong question being addressed.

We concentrate on problems where Ω_θ is a subset of R^d, i.e., d-dimensional real space. These are so-called *fully parametric* problems. Other possibilities are to have semiparametric problems or fully nonparametric problems. These typically involve fewer assumptions of structure and distributional form but usually contain strong assumptions about independencies. To an appreciable extent the formal theory of semiparametric models aims to parallel that of parametric models.

The probability model and the choice of ψ serve to translate a subject-matter question into a mathematical and statistical one and clearly the faithfulness of the translation is crucial. To check on the appropriateness of a new type of model to represent a data-generating process it is sometimes helpful to consider how the model could be used to generate synthetic data. This is especially the case for stochastic process models. Understanding of new or unfamiliar models can be obtained both by mathematical analysis and by simulation, exploiting the power of modern computational techniques to assess the kind of data generated by a specific kind of model.

1.2 Role of formal theory of inference

The formal theory of inference initially takes the family of models as given and the objective as being to answer questions about the model in the light of the data. Choice of the family of models is, as already remarked, obviously crucial but outside the scope of the present discussion. More than one choice may be needed to answer different questions.

A second and complementary phase of the theory concerns what is sometimes called *model criticism*, addressing whether the data suggest minor or major modification of the model or in extreme cases whether the whole focus of the analysis should be changed. While model criticism is often done rather informally in practice, it is important for any formal theory of inference that it embraces the issues involved in such checking.

1.3 Some simple models

General notation is often not best suited to special cases and so we use more conventional notation where appropriate.

Example 1.1. *The normal mean.* Whenever it is required to illustrate some point in simplest form it is almost inevitable to return to the most hackneyed of examples, which is therefore given first. Suppose that $Y_1, \ldots, Y_n$ are independently normally distributed with unknown mean μ and known variance σ_0^2. Here μ plays the role of the unknown parameter θ in the general formulation. In one of many possible generalizations, the variance σ^2 also is unknown. The parameter vector is then (μ, σ^2). The component of interest ψ would often be μ

but could be, for example, σ^2 or μ/σ, depending on the focus of subject-matter interest.

Example 1.2. *Linear regression.* Here the data are n pairs $(y_1, z_1), \ldots, (y_n, z_n)$ and the model is that $Y_1, \ldots, Y_n$ are independently normally distributed with variance σ^2 and with

$$E(Y_k) = \alpha + \beta z_k. \tag{1.3}$$

Here typically, but not necessarily, the parameter of interest is $\psi = \beta$ and the nuisance parameter is $\lambda = (\alpha, \sigma^2)$. Other possible parameters of interest include the intercept at $z = 0$, namely α, and $-\alpha/\beta$, the intercept of the regression line on the z-axis.

Example 1.3. *Linear regression in semiparametric form.* In Example 1.2 replace the assumption of normality by an assumption that the Y_k are uncorrelated with constant variance. This is semiparametric in that the systematic part of the variation, the linear dependence on z_k, is specified parametrically and the random part is specified only via its covariance matrix, leaving the functional form of its distribution open. A complementary form would leave the systematic part of the variation a largely arbitrary function and specify the distribution of error parametrically, possibly of the same normal form as in Example 1.2. This would lead to a discussion of smoothing techniques.

Example 1.4. *Linear model.* We have an $n \times 1$ vector Y and an $n \times q$ matrix z of fixed constants such that

$$E(Y) = z\beta, \quad \text{cov}(Y) = \sigma^2 I, \tag{1.4}$$

where β is a $q \times 1$ vector of unknown parameters, I is the $n \times n$ identity matrix and with, in the analogue of Example 1.2, the components independently normally distributed. Here z is, in initial discussion at least, assumed of full rank $q < n$. A relatively simple but important generalization has $\text{cov}(Y) = \sigma^2 V$, where V is a given positive definite matrix. There is a corresponding semiparametric version generalizing Example 1.3.

Both Examples 1.1 and 1.2 are special cases, in the former the matrix z consisting of a column of 1s.

Example 1.5. *Normal-theory nonlinear regression.* Of the many generalizations of Examples 1.2 and 1.4, one important possibility is that the dependence on the parameters specifying the systematic part of the structure is nonlinear. For example, instead of the linear regression of Example 1.2 we might wish to consider

$$E(Y_k) = \alpha + \beta \exp(\gamma z_k), \tag{1.5}$$

where from the viewpoint of statistical theory the important nonlinearity is not in the dependence on the variable z but rather that on the parameter γ.

More generally the equation $E(Y) = z\beta$ in (1.4) may be replaced by

$$E(Y) = \mu(\beta), \qquad (1.6)$$

where the $n \times 1$ vector $\mu(\beta)$ is in general a nonlinear function of the unknown parameter β and also of the explanatory variables.

Example 1.6. *Exponential distribution.* Here the data are $(y_1, \ldots, y_n)$ and the model takes $Y_1, \ldots, Y_n$ to be independently exponentially distributed with density $\rho e^{-\rho y}$, for $y > 0$, where $\rho > 0$ is an unknown rate parameter. Note that possible parameters of interest are ρ, $\log \rho$ and $1/\rho$ and the issue will arise of possible invariance or equivariance of the inference under reparameterization, i.e., shifts from, say, ρ to $1/\rho$. The observations might be intervals between successive points in a Poisson process of rate ρ. The interpretation of $1/\rho$ is then as a mean interval between successive points in the Poisson process. The use of $\log \rho$ would be natural were ρ to be decomposed into a product of effects of different explanatory variables and in particular if the ratio of two rates were of interest.

Example 1.7. *Comparison of binomial probabilities.* Suppose that the data are (r_0, n_0) and (r_1, n_1), where r_k denotes the number of successes in n_k binary trials under condition k. The simplest model is that the trials are mutually independent with probabilities of success π_0 and π_1. Then the random variables R_0 and R_1 have independent binomial distributions. We want to compare the probabilities and for this may take various forms for the parameter of interest, for example

$$\psi = \log\{\pi_1/(1-\pi_1)\} - \log\{\pi_0/(1-\pi_0)\}, \quad \text{or} \quad \psi = \pi_1 - \pi_0, \qquad (1.7)$$

and so on. For many purposes it is immaterial how we define the complementary parameter λ. Interest in the nonlinear function $\log\{\pi/(1-\pi)\}$ of a probability π stems partly from the interpretation as a log odds, partly because it maps the parameter space $(0, 1)$ onto the real line and partly from the simplicity of some resulting mathematical models of more complicated dependences, for example on a number of explanatory variables.

Example 1.8. *Location and related problems.* A different generalization of Example 1.1 is to suppose that $Y_1, \ldots, Y_n$ are independently distributed all with the density $g(y - \mu)$, where $g(y)$ is a given probability density. We call μ

a *location parameter*; often it may by convention be taken to be the mean or median of the density.

A further generalization is to densities of the form $\tau^{-1}g\{(y-\mu)/\tau\}$, where τ is a positive parameter called a *scale parameter* and the family of distributions is called a location and scale family.

Central to the general discussion of such models is the notion of a family of transformations of the underlying random variable and the parameters. In the location and scale family if Y_k is transformed to $aY_k + b$, where $a > 0$ and b are arbitrary, then the new random variable has a distribution of the original form with transformed parameter values

$$a\mu + b, \ a\tau. \tag{1.8}$$

The implication for most purposes is that any method of analysis should obey the same transformation properties. That is, if the limits of uncertainty for say μ, based on the original data, are centred on $\tilde{y}$, then the limits of uncertainty for the corresponding parameter after transformation are centred on $a\tilde{y} + b$.

Typically this represents, in particular, the notion that conclusions should not depend on the units of measurement. Of course, some care is needed with this idea. If the observations are temperatures, for some purposes arbitrary changes of scale and location, i.e., of the nominal zero of temperature, are allowable, whereas for others recognition of the absolute zero of temperature is essential. In the latter case only transformations from kelvins to some multiple of kelvins would be acceptable.

It is sometimes important to distinguish invariance that springs from some subject-matter convention, such as the choice of units of measurement from invariance arising out of some mathematical formalism.

The idea underlying the above example can be expressed in much more general form involving two groups of transformations, one on the sample space and one on the parameter space. Data recorded as directions of vectors on a circle or sphere provide one example. Another example is that some of the techniques of normal-theory multivariate analysis are invariant under arbitrary nonsingular linear transformations of the observed vector, whereas other methods, notably principal component analysis, are invariant only under orthogonal transformations.

The object of the study of a theory of statistical inference is to provide a set of ideas that deal systematically with the above relatively simple situations and, more importantly still, enable us to deal with new models that arise in new applications.

1.4 Formulation of objectives

We can, as already noted, formulate possible objectives in two parts as follows. Part I takes the family of models as given and aims to:

- give intervals or in general sets of values within which ψ is in some sense likely to lie;
- assess the consistency of the data with a particular parameter value ψ_0;
- predict as yet unobserved random variables from the same random system that generated the data;
- use the data to choose one of a given set of decisions $\mathcal{D}$, requiring the specification of the consequences of various decisions.

Part II uses the data to examine the family of models via a process of model criticism. We return to this issue in Section 3.2.

We shall concentrate in this book largely but not entirely on the first two of the objectives in Part I, interval estimation and measuring consistency with specified values of ψ.

To an appreciable extent the theory of inference is concerned with generalizing to a wide class of models two approaches to these issues which will be outlined in the next section and with a critical assessment of these approaches.

1.5 Two broad approaches to statistical inference

1.5.1 General remarks

Consider the first objective above, that of providing intervals or sets of values likely in some sense to contain the parameter of interest, ψ.

There are two broad approaches, called *frequentist* and *Bayesian*, respectively, both with variants. Alternatively the former approach may be said to be based on *sampling theory* and an older term for the latter is that it uses *inverse probability*. Much of the rest of the book is concerned with the similarities and differences between these two approaches. As a prelude to the general development we show a very simple example of the arguments involved.

We take for illustration Example 1.1, which concerns a normal distribution with unknown mean μ and known variance. In the formulation probability is used to model variability as experienced in the phenomenon under study and its meaning is as a long-run frequency in repetitions, possibly, or indeed often, hypothetical, of that phenomenon.

What can reasonably be said about μ on the basis of observations $y_1, \ldots, y_n$ and the assumptions about the model?

1.5.2 Frequentist discussion

In the first approach we make no further probabilistic assumptions. In particular we treat μ as an unknown constant. Strong arguments can be produced for reducing the data to their mean $\bar{y} = \Sigma y_k/n$, which is the observed value of the corresponding random variable $\bar{Y}$. This random variable has under the assumptions of the model a normal distribution of mean μ and variance σ_0^2/n, so that in particular

$$P(\bar{Y} > \mu - k_c^* \sigma_0/\sqrt{n}) = 1 - c, \qquad (1.9)$$

where, with $\Phi(.)$ denoting the standard normal integral, $\Phi(k_c^*) = 1 - c$. For example with $c = 0.025, k_c^* = 1.96$. For a sketch of the proof, see Note 1.5.

Thus the statement equivalent to (1.9) that

$$P(\mu < \bar{Y} + k_c^* \sigma_0/\sqrt{n}) = 1 - c, \qquad (1.10)$$

can be interpreted as specifying a hypothetical long run of statements about μ a proportion $1 - c$ of which are correct. We have observed the value $\bar{y}$ of the random variable $\bar{Y}$ and the statement

$$\mu < \bar{y} + k_c^* \sigma_0/\sqrt{n} \qquad (1.11)$$

is thus one of this long run of statements, a specified proportion of which are correct. In the most direct formulation of this μ is fixed and the statements vary and this distinguishes the statement from a probability distribution for μ. In fact a similar interpretation holds if the repetitions concern an arbitrary sequence of fixed values of the mean.

There are a large number of generalizations of this result, many underpinning standard elementary statistical techniques. For instance, if the variance σ^2 is unknown and estimated by $\Sigma(y_k - \bar{y})^2/(n - 1)$ in (1.9), then k_c^* is replaced by the corresponding point in the Student t distribution with $n - 1$ degrees of freedom.

There is no need to restrict the analysis to a single level c and provided concordant procedures are used at the different c a formal distribution is built up.

Arguments involving probability only via its (hypothetical) long-run frequency interpretation are called *frequentist*. That is, we define procedures for assessing evidence that are calibrated by how they would perform were they used repeatedly. In that sense they do not differ from other measuring instruments. We intend, of course, that this long-run behaviour is some assurance that with our particular data currently under analysis sound conclusions are drawn. This raises important issues of ensuring, as far as is feasible, the relevance of the long run to the specific instance.

1.5.3 Bayesian discussion

In the second approach to the problem we treat μ as having a probability distribution both with and without the data. This raises two questions: what is the meaning of probability in such a context, some extended or modified notion of probability usually being involved, and how do we obtain numerical values for the relevant probabilities? This is discussed further later, especially in Chapter 5. For the moment we assume some such notion of probability concerned with measuring uncertainty is available.

If indeed we can treat μ as the realized but unobserved value of a random variable M, all is in principle straightforward. By Bayes' theorem, i.e., by simple laws of probability,

$$f_{M|Y}(\mu \mid y) = f_{Y|M}(y \mid \mu) f_M(\mu) \Big/ \int f_{Y|M}(y \mid \phi) f_M(\phi) d\phi. \qquad (1.12)$$

The left-hand side is called the *posterior density* of M and of the two terms in the numerator the first is determined by the model and the other, $f_M(\mu)$, forms the *prior distribution* summarizing information about M not arising from y. Any method of inference treating the unknown parameter as having a probability distribution is called *Bayesian* or, in an older terminology, an argument of *inverse probability*. The latter name arises from the inversion of the order of target and conditioning events as between the model and the posterior density.

The intuitive idea is that in such cases all relevant information about μ is then contained in the conditional distribution of the parameter given the data, that this is determined by the elementary formulae of probability theory and that remaining problems are solely computational.

In our example suppose that the prior for μ is normal with known mean m and variance v. Then the posterior density for μ is proportional to

$$\exp\{-\Sigma(y_k - \mu)^2/(2\sigma_0^2) - (\mu - m)^2/(2v)\} \qquad (1.13)$$

considered as a function of μ. On completing the square as a function of μ, there results a normal distribution of mean and variance respectively

$$\frac{\bar{y}/(\sigma_0^2/n) + m/v}{1/(\sigma_0^2/n) + 1/v}, \qquad (1.14)$$

$$\frac{1}{1/(\sigma_0^2/n) + 1/v}; \qquad (1.15)$$

for more details of the argument, see Note 1.5. Thus an upper limit for μ satisfied with posterior probability $1 - c$ is

$$\frac{\bar{y}/(\sigma_0^2/n) + m/v}{1/(\sigma_0^2/n) + 1/v} + k_c^* \sqrt{\frac{1}{1/(\sigma_0^2/n) + 1/v}}. \qquad (1.16)$$

If v is large compared with σ_0^2/n and m is not very different from $\bar{y}$ these limits agree closely with those obtained by the frequentist method. If there is a serious discrepancy between $\bar{y}$ and m this indicates either a flaw in the data or a misspecification of the prior distribution.

This broad parallel between the two types of analysis is in no way specific to the normal distribution.

1.6 Some further discussion

We now give some more detailed discussion especially of Example 1.4 and outline a number of special models that illustrate important issues.

The linear model of Example 1.4 and methods of analysis of it stemming from the method of least squares are of much direct importance and also are the base of many generalizations. The central results can be expressed in matrix form centring on the least squares estimating equations

$$z^T z \hat{\beta} = z^T Y, \tag{1.17}$$

the vector of fitted values

$$\hat{Y} = z\hat{\beta}, \tag{1.18}$$

and the residual sum of squares

$$\text{RSS} = (Y - \hat{Y})^T (Y - \hat{Y}) = Y^T Y - \hat{\beta}^T (z^T z) \hat{\beta}. \tag{1.19}$$

Insight into the form of these results is obtained by noting that were it not for random error the vector Y would lie in the space spanned by the columns of z, that $\hat{Y}$ is the orthogonal projection of Y onto that space, defined thus by

$$z^T (Y - \hat{Y}) = z^T (Y - z\hat{\beta}) = 0 \tag{1.20}$$

and that the residual sum of squares is the squared norm of the component of Y orthogonal to the columns of z. See Figure 1.1.

There is a fairly direct generalization of these results to the nonlinear regression model of Example 1.5. Here if there were no error the observations would lie on the surface defined by the vector $\mu(\beta)$ as β varies. Orthogonal projection involves finding the point $\mu(\hat{\beta})$ closest to Y in the least squares sense, i.e., minimizing the sum of squares of deviations $\{Y - \mu(\beta)\}^T \{Y - \mu(\beta)\}$. The resulting equations defining $\hat{\beta}$ are best expressed by defining

$$z^T(\beta) = \nabla \mu^T(\beta), \tag{1.21}$$

where ∇ is the $q \times 1$ gradient operator with respect to β, i.e., $\nabla^T = (\partial/\partial \beta_1, \ldots, \partial/\partial \beta_q)$. Thus $z(\beta)$ is an $n \times q$ matrix, reducing to the previous z

1.6 Some further discussion

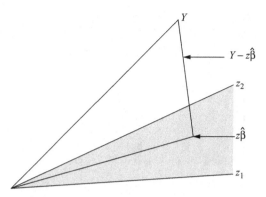

Figure 1.1. Linear model. Without random error the vector Y would lie in the shaded space spanned by the columns $z_1, z_2, \ldots$ of the matrix z. The least squares estimate $\hat{\beta}$ is defined by orthogonal projection of Y onto the space determined by z. For orthogonality the vector $Y - z\hat{\beta}$ must be orthogonal to the vectors $z_1, z_2, \ldots$. Further a Pythogorean identity holds for the squared length $Y^T Y$ decomposing it into the residual sum of squares, RSS, and the squared length of the vector of fitted values.

in the linear case. Just as the columns of z define the linear model, the columns of $z(\beta)$ define the tangent space to the model surface evaluated at β. The least squares estimating equation is thus

$$z^T(\hat{\beta})\{Y - \mu(\hat{\beta})\} = 0. \tag{1.22}$$

The local linearization implicit in this is valuable for numerical iteration. One of the simplest special cases arises when $E(Y_k) = \beta_0 \exp(-\beta_1 z_k)$ and the geometry underlying the nonlinear least squares equations is summarized in Figure 1.2.

The simple examples used here in illustration have one component random variable attached to each observation and all random variables are mutually independent. In many situations random variation comes from several sources and random components attached to different component observations may not be independent, showing for example temporal or spatial dependence.

Example 1.9. *A component of variance model.* The simplest model with two sources of random variability is the normal-theory component of variance formulation, which for random variables $Y_{ks}; k = 1, \ldots, m; s = 1, \ldots, t$ has the form

$$Y_{ks} = \mu + \eta_k + \epsilon_{ks}. \tag{1.23}$$

Here μ is an unknown mean and the η and the ϵ are mutually independent normally distributed random variables with zero mean and variances respectively

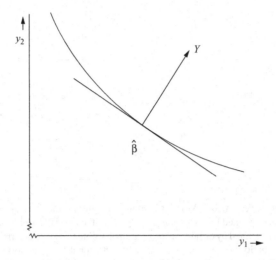

Figure 1.2. Nonlinear least squares estimation. Without random error the observations would lie on the curved surface shown here for just two observations. The least squares estimates are obtained by projection such that the residual vector is orthogonal to the tangent plane to the surface at the fitted point.

τ_η, τ_ϵ, called *components of variance*. This model represents the simplest form for the resolution of random variability into two components. The model could be specified by the simple block-diagonal form of the covariance matrix of Y considered as a single column vector.

Models of this broad type with several layers of variation are sometimes called Bayesian, a misleading terminology that will not be used here.

Example 1.10. *Markov models.* For any sequence of random variables $Y_1, \ldots, Y_n$ the joint density can be factorized recursively in the form

$$f_{Y_1}(y_1) f_{Y_2|Y_1}(y_2; y_1) \ldots f_{Y_n|Y_1,\ldots,Y_{n-1}}(y_n; y_1, \ldots, y_{n-1}). \qquad (1.24)$$

If the process is a Markov process in which very often the sequence is in time, there is the major simplification that in each term the conditioning is only on the preceding term, so that the density is

$$f_{Y_1}(y_1) \Pi f_{Y_k|Y_{k-1}}(y_k; y_{k-1}). \qquad (1.25)$$

That is, to produce a parametric Markov process we have to specify only the one-step transition probabilities in parametric form.

Another commonly occurring more complicated form of variation arises with time series, spatial and spatial-temporal data. The simplest time series model is the Gaussian first-order autoregression, a Markov process defined by the equation

$$Y_t - \mu = \beta(Y_{t-1} - \mu) + \epsilon_t. \tag{1.26}$$

Here the forcing terms, ϵ_t, are called *innovations*. They are assumed to be independent and identically distributed random variables of zero mean and variance σ_ϵ^2. The specification is completed once the value or distribution of the initial condition Y_1 is set out.

Extensions of this idea to spatial and spatial-temporal data are important; see Note 1.6.

1.7 Parameters

A central role is played throughout the book by the notion of a parameter vector, θ. Initially this serves to index the different probability distributions making up the full model. If interest were exclusively in these probability distributions as such, any $(1, 1)$ transformation of θ would serve equally well and the choice of a particular version would be essentially one of convenience. For most of the applications in mind here, however, the interpretation is via specific parameters and this raises the need both to separate parameters of interest, ψ, from nuisance parameters, λ, and to choose specific representations. In relatively complicated problems where several different research questions are under study different parameterizations may be needed for different purposes.

There are a number of criteria that may be used to define the individual component parameters. These include the following:

- the components should have clear subject-matter interpretations, for example as differences, rates of change or as properties such as in a physical context mass, energy and so on. If not dimensionless they should be measured on a scale unlikely to produce very large or very small values;
- it is desirable that this interpretation is retained under reasonable perturbations of the model;
- different components should not have highly correlated errors of estimation;
- statistical theory for estimation should be simple;
- if iterative methods of computation are needed then speedy and assured convergence is desirable.

The first criterion is of primary importance for parameters of interest, at least in the presentation of conclusions, but for nuisance parameters the other criteria are of main interest. There are considerable advantages in formulations leading to simple methods of analysis and judicious simplicity is a powerful aid to understanding, but for parameters of interest subject-matter meaning must have priority.

Notes 1

General. There are quite a number of books setting out in some detail the mathematical details for this chapter and for much of the subsequent material. Davison (2003) provides an extended and wide-ranging account of statistical methods and their theory. Good introductory accounts from a more theoretical stance are Rice (1988) and, at a somewhat more advanced level, Casella and Berger (1990). For an introduction to Bayesian methods, see Box and Tiao (1973). Rao (1973) gives a wide-ranging account of statistical theory together with some of the necessary mathematical background. Azzalini (1996), Pawitan (2000) and Severini (2001) emphasize the role of likelihood, as forms the basis of the present book. Silvey (1970) gives a compact and elegant account of statistical theory taking a different viewpoint from that adopted here. Young and Smith (2005) give a broad introduction to the theory of statistical inference. The account of the theory of statistics by Cox and Hinkley (1974) is more detailed than that in the present book. Barnett and Barnett (1999) give a broad comparative view. Williams (2001) provides an original and wide-ranging introduction to probability and to statistical theory.

Section 1.1. The sequence question–data–analysis is emphasized here. In *data mining* the sequence may be closer to data–analysis–question. Virtually the same statistical methods may turn out to be useful, but the probabilistic interpretation needs considerable additional caution in the latter case.

More formally (1.2) can be written

$$E(Y) = \int y f_Y(y; \theta) \nu(dy),$$

where $\nu(.)$ is a dominating measure, typically to be thought of as dy in regions of continuity and having atoms of unit size where Y has a positive probability. The support of a distribution is loosely the set of points with nonzero density; more formally every open set containing a point of support should have positive probability. In most but not all problems the support does not depend on θ.

A great deal can be achieved from the properties of E and the associated notions of variance and covariance, i.e., from first and second moment calculations. In particular, subject only to existence the expected value of a sum is the sum of the expected values and for independent random variables the expectation of a product is the product of the expectations. There is also a valuable property of conditional expectations summarized in $E(Y) = E_X E(Y \mid X)$. An important extension obtained by applying this last property also to Y^2 is that $\text{var}(Y) = E_X\{\text{var}(Y \mid X)\} + \text{var}_X\{E(Y \mid X)\}$. That is, an unconditional variance is the expectation of a conditional variance plus the variance of a conditional mean.

The random variables considered all take values in some finite-dimensional real space. It is possible to consider more general possibilities, such as complex-valued random variables or random variables in some space of infinite dimension, such as Hilbert space, but this will not be necessary here. The so-called singular continuous distributions of probability theory arise only as absolutely continuous distributions on some manifold of lower dimension than the space originally considered. In general discussions of probability, the density for absolutely continuous random variables is not uniquely defined, only integrals over sets being meaningful. This nonuniqueness is of no concern here, partly because all real distributions are discrete and partly because there is nothing to be gained by considering versions of the density other than those well behaved for all members of the family of distributions under study.

Section 1.3. The examples in this section are mostly those arising in accounts of elementary statistical methods. All of them can be generalized in various ways. The assumptions of Example 1.3 form the basis of the discussion of least squares theory by Gauss leading to the notion of linear unbiased estimates of minimum variance.

Section 1.4. It may seem odd to exclude point estimation from the list of topics but it is most naturally regarded either as a decision problem, involving therefore no explicit statement of uncertainty, or as an intermediate stage of analysis of relatively complex data.

Section 1.5. The moment generating function $M_Y(p)$ for a random variable Y, equivalent except for sign to a Laplace transform of the density, is $E(e^{pY})$ or in the vector case $E(e^{p^T Y})$. For a normal distribution of mean μ and variance σ^2 it is $\exp(p\mu + p^2\sigma^2/2)$. For the sum of independent random variables, $S = \Sigma Y_k$ the product law of expectations gives $M_S(p) = \exp(\Sigma \mu_k p + \Sigma \sigma_k^2 p^2/2)$. A uniqueness theorem then proves that S and hence $\bar{Y}$ has a normal distribution.

Equation (1.12) is the version for densities of the immediate consequence of the definition of conditional probability. For events A and B this takes the form

$$P(B \mid A) = P(A \cap B)/P(A) = P(A \mid B)P(B)/P(A).$$

The details of the calculation for (1.13), (1.14) and (1.15) have been omitted. Note that $\Sigma(y_k - \mu)^2 = n(\bar{y} - \mu)^2 + \Sigma(y_k - \bar{y})^2$, in which the second term can be ignored because it does not involve μ, and that finally as a function of μ we can write $A\mu^2 + 2B\mu + C$ as $A(\mu + B/A)^2$ plus a constant independent of μ.

Section 1.6. Most of the general books mentioned above discuss the normal-theory linear model as do the numerous textbooks on regression analysis. For a systematic study of components of variance, see, for instance, Cox and Solomon (2002) and for time series analysis Brockwell and Davis (1991, 1998). For spatial processes, see Besag (1974) and Besag and Mondal (2005).

Section 1.7. For a discussion of choice of parameterization especially in nonlinear models, see Ross (1990).

2
Some concepts and simple applications

Summary. An introduction is given to the notions of likelihood and sufficiency and the exponential family of distributions is defined and exemplified.

2.1 Likelihood

The *likelihood* for the vector of observations y is defined as

$$\text{lik}(\theta; y) = f_Y(y; \theta), \tag{2.1}$$

considered in the first place as a function of θ for given y. Mostly we work with its logarithm $l(\theta; y)$, often abbreviated to $l(\theta)$. Sometimes this is treated as a function of the random vector Y rather than of y. The log form is convenient, in particular because f will often be a product of component terms. Occasionally we work directly with the likelihood function itself. For nearly all purposes multiplying the likelihood formally by an arbitrary function of y, or equivalently adding an arbitrary such function to the log likelihood, would leave unchanged that part of the analysis hinging on direct calculations with the likelihood.

Any calculation of a posterior density, whatever the prior distribution, uses the data only via the likelihood. Beyond that, there is some intuitive appeal in the idea that differences in $l(\theta)$ measure the relative effectiveness of different parameter values θ in explaining the data. This is sometimes elevated into a principle called *the law of the likelihood.*

A key issue concerns the additional arguments needed to extract useful information from the likelihood, especially in relatively complicated problems possibly with many nuisance parameters. Likelihood will play a central role in almost all the arguments that follow.

2.2 Sufficiency

2.2.1 Definition

The term *statistic* is often used (rather oddly) to mean any function of the observed random variable Y or its observed counterpart. A statistic $S = S(Y)$ is called *sufficient* under the model if the conditional distribution of Y given $S = s$ is independent of θ for all s, θ. Equivalently

$$l(\theta; y) = \log h(s, \theta) + \log m(y), \tag{2.2}$$

for suitable functions h and m. The equivalence forms what is called the Neyman factorization theorem. The proof in the discrete case follows most explicitly by defining any new variable W, a function of Y, such that Y is in $(1,1)$ correspondence with (S, W), i.e., such that (S, W) determines Y. The individual atoms of probability are unchanged by transformation. That is,

$$f_Y(y; \theta) = f_{S,W}(s, w; \theta) = f_S(s; \theta) f_{W|S}(w; s), \tag{2.3}$$

where the last term is independent of θ by definition. In the continuous case there is the minor modification that a Jacobian, not involving θ, is needed when transforming from Y to (S, W). See Note 2.2.

We use the minimal form of S; i.e., extra components could always be added to any given S and the sufficiency property retained. Such addition is undesirable and is excluded by the requirement of minimality. The minimal form always exists and is essentially unique.

Any Bayesian inference uses the data only via the minimal sufficient statistic. This is because the calculation of the posterior distribution involves multiplying the likelihood by the prior and normalizing. Any factor of the likelihood that is a function of y alone will disappear after normalization.

In a broader context the importance of sufficiency can be considered to arise as follows. Suppose that instead of observing $Y = y$ we were equivalently to be given the data in two stages:

- first we observe $S = s$, an observation from the density $f_S(s; \theta)$;
- then we are given the remaining data, in effect an observation from the density $f_{Y|S}(y; s)$.

Now, so long as the model holds, the second stage is an observation on a fixed and known distribution which could as well have been obtained from a random number generator. Therefore $S = s$ contains all the information about θ given the model, whereas the conditional distribution of Y given $S = s$ allows assessment of the model.

2.2 Sufficiency

There are thus two implications of sufficiency. One is that given the model and disregarding considerations of robustness or computational simplicity the data should be used only via s. The second is that the *known* distribution $f_{Y|S}(y \mid s)$ is available to examine the adequacy of the model. Illustrations will be given in Chapter 3.

Sufficiency may reduce the dimensionality of a problem but we still have to determine what to do with the sufficient statistic once found.

2.2.2 Simple examples

Example 2.1. *Exponential distribution (ctd).* The likelihood for Example 1.6 is

$$\rho^n \exp(-\rho \Sigma y_k), \tag{2.4}$$

so that the log likelihood is

$$n \log \rho - \rho \Sigma y_k, \tag{2.5}$$

and, assuming n to be fixed, involves the data only via Σy_k or equivalently via $\bar{y} = \Sigma y_k/n$. By the factorization theorem the sum (or mean) is therefore sufficient. Note that had the sample size also been random the sufficient statistic would have been $(n, \Sigma y_k)$; see Example 2.4 for further discussion.

In this example the density of $S = \Sigma Y_k$ is $\rho(\rho s)^{n-1} e^{-\rho s}/(n-1)!$, a gamma distribution. It follows that ρS has a fixed distribution. It follows also that the joint conditional density of the Y_k given $S = s$ is uniform over the simplex $0 \leq y_k \leq s; \Sigma y_k = s$. This can be used to test the adequacy of the model.

Example 2.2. *Linear model (ctd).* A minimal sufficient statistic for the linear model, Example 1.4, consists of the least squares estimates and the residual sum of squares. This strong justification of the use of least squares estimates depends on writing the log likelihood in the form

$$-n \log \sigma - (y - z\beta)^T (y - z\beta)/(2\sigma^2) \tag{2.6}$$

and then noting that

$$(y - z\beta)^T (y - z\beta) = (y - z\hat{\beta})^T (y - z\hat{\beta}) + (\hat{\beta} - \beta)^T (z^T z)(\hat{\beta} - \beta), \tag{2.7}$$

in virtue of the equations defining the least squares estimates. This last identity has a direct geometrical interpretation. The squared norm of the vector defined by the difference between Y and its expected value $z\beta$ is decomposed into a component defined by the difference between Y and the estimated mean $z\hat{\beta}$ and an orthogonal component defined via $\hat{\beta} - \beta$. See Figure 1.1.

It follows that the log likelihood involves the data only via the least squares estimates and the residual sum of squares. Moreover, if the variance σ^2 were known, the residual sum of squares would be a constant term in the log likelihood and hence the sufficient statistic would be reduced to $\hat{\beta}$ alone.

This argument fails for a regression model nonlinear in the parameters, such as the exponential regression (1.5). In the absence of error the $n \times 1$ vector of observations then lies on a curved surface and while the least squares estimates are still given by orthogonal projection they satisfy nonlinear equations and the decomposition of the log likelihood which is the basis of the argument for sufficiency holds only as an approximation obtained by treating the curved surface as locally flat.

Example 2.3. *Uniform distribution.* Suppose that $Y_1, \ldots, Y_n$ are independent and identically distributed in a uniform distribution on (θ_1, θ_2). Let $Y_{(1)}, Y_{(n)}$ be the smallest and largest order statistics, i.e., the random variables $\min(Y_1, \ldots, Y_n)$ and $\max(Y_1, \ldots, Y_n)$ respectively. Then the likelihood involves the data only via $(y_{(1)}, y_{(n)})$, being zero unless the interval $(y_{(1)}, y_{(n)})$ is within (θ_1, θ_2). Therefore the two order statistics are minimal sufficient. Similar results hold when one terminus is known or when the range of the distribution, i.e., $\theta_2 - \theta_1$, is known.

In general, care is needed in writing down the likelihood and in evaluating the resulting procedures whenever the form of the likelihood has discontinuities, in particular, as here, where the support of the probability distribution depends on the unknown parameter.

Example 2.4. *Binary fission.* Suppose that in a system of particles each particle has in small time period $(t, t + \Delta)$ a probability $\rho \Delta + o(\Delta)$ of splitting into two particles with identical properties and a probability $1 - \rho \Delta + o(\Delta)$ of remaining unchanged, events for different particles being independent. Suppose that the system is observed for a time period $(0, t_0)$. Then by dividing the whole exposure history into small intervals separately for each individual we have that the likelihood is

$$\rho^n e^{-\rho t_e}. \tag{2.8}$$

Here n is the number of new particles born and t_e is the total exposure time, i.e., the sum of all life-times of individuals in the system, and both are random variables, so that the sufficient statistic is $S = (N, T_e)$.

2.3 Exponential family

We now consider a family of models that are important both because they have relatively simple properties and because they capture in one discussion many, although by no means all, of the inferential properties connected with important standard distributions, the binomial, Poisson and geometric distributions and the normal and gamma distributions, and others too.

Suppose that θ takes values in a well-behaved region in R^d, and not of lower dimension, and that we can find a $(d \times 1)$-dimensional statistic s and a parameterization ϕ, i.e., a $(1,1)$ transformation of θ, such that the model has the form

$$m(y)\exp\{s^T\phi - k(\phi)\}, \qquad (2.9)$$

where $s = s(y)$ is a function of the data. Then S is sufficient; subject to some important regularity conditions this is called a *regular or full* (d,d) *exponential family* of distributions. The statistic S is called the *canonical statistic* and ϕ the *canonical parameter*. The parameter $\eta = E(S; \phi)$ is called the *mean parameter*. Because the stated function defines a distribution, it follows that

$$\int m(y)\exp\{s^T\phi - k(\phi)\}dy = 1, \qquad (2.10)$$

$$\int m(y)\exp\{s^T(\phi+p) - k(\phi+p)\}dy = 1. \qquad (2.11)$$

Hence, noting that $s^T p = p^T s$, we have from (2.11) that the moment generating function of S is

$$E\{\exp(p^T S)\} = \exp\{k(\phi+p) - k(\phi)\}. \qquad (2.12)$$

Therefore the cumulant generating function of S, defined as the log of the moment generating function, is

$$k(\phi+p) - k(\phi), \qquad (2.13)$$

providing a fairly direct interpretation of the function $k(.)$. Because the mean is given by the first derivatives of that generating function, we have that $\eta = \nabla k(\phi)$, where ∇ is the gradient operator $(\partial/\partial\phi_1, \ldots, \partial/\partial\phi_d)^T$. See Note 2.3 for a brief account of both cumulant generating functions and the ∇ notation.

Example 2.5. *Binomial distribution.* If R denotes the number of successes in n independent binary trials each with probability of success π, its density can be written

$$n!/\{r!(n-r)!\}\pi^r(1-\pi)^{n-r} = m(r)\exp\{r\phi - n\log(1+e^\phi)\}, \qquad (2.14)$$

say, where $\phi = \log\{\pi/(1-\pi)\}$, often called the log odds, is the canonical parameter and r the canonical statistic. Note that the mean parameter is $E(R) = n\pi$ and can be recovered also by differentiating $k(\phi)$.

In general for continuous distributions a fixed smooth (1,1) transformation of the random variable preserves the exponential family form; the Jacobian of the transformation changes only the function $m(y)$. Thus if Y has the exponential distribution $\rho e^{-\rho y}$, or in exponential family form $\exp(-\rho y + \log \rho)$, then $X = \sqrt{Y}$ has the Rayleigh distribution $2\rho x \exp(-\rho x^2)$, or in exponential family form $2x \exp(-\rho x^2 + \log \rho)$. The effect of the transformation from Y to $\sqrt{Y}$ is thus to change the initializing function $m(.)$ but otherwise leave the canonical statistic taking the same value and leaving the canonical parameter and the function $k(.)$ unchanged.

In general the likelihood as a function of ϕ is maximized by solving the equation in $\hat{\phi}$

$$s = \nabla k(\phi)_{\phi=\hat{\phi}} = \hat{\eta}, \qquad (2.15)$$

where it can be shown that for a regular exponential family the solution is unique and indeed corresponds to a maximum, unless possibly if s lies on the boundary of its support. We call $\hat{\phi}$ and $\hat{\eta}$ the *maximum likelihood estimates* of ϕ and η respectively. In (2.15) we equate the canonical statistic to its expectation to form an estimating equation which has a direct appeal, quite apart from more formal issues connected with likelihood. The important role of the maximum likelihood estimate will appear later.

A further notion is that of a curved exponential family. Suppose that we start with a full exponential family in which the canonical parameter ϕ and canonical statistic s are both of dimension k. Now suppose that ϕ is restricted to lie in a curved space of dimension d, where $d < k$. Then the exponential family form will still hold but now the $k \times 1$ vector ϕ is a function of d originating parameters defining position in the curved space. This is called a (k,d) *curved exponential family*. The statistical properties of curved exponential families are much less simple than those of full exponential families. The normal-theory nonlinear regression already discussed is an important example. The following are simpler instances.

Example 2.6. *Fisher's hyperbola.* Suppose that the model is that the pairs (Y_{k1}, Y_{k2}) are independently normally distributed with unit variance and with means $(\theta, c/\theta)$, where c is a known constant. The log likelihood for a single observation is

$$y_{k1}\theta + cy_{k2}/\theta - \theta^2/2 - c^2\theta^{-2}/2. \qquad (2.16)$$

Note that the squared terms in the normal density may be omitted because they are not functions of the unknown parameter. It follows that for the full set of observations the sample totals or means form the minimal sufficient

2.4 Choice of priors for exponential family problems

statistic. Thus what in the full family would be the two-dimensional parameter (ϕ_1, ϕ_2) is confined to the hyperbola $(\theta, c/\theta)$ forming a (2, 1) family with a two-dimensional sufficient statistic and a one-dimensional parameter.

Example 2.7. *Binary fission (ctd).* As shown in Example 2.4 the likelihood is a function of n, the number of births, and t_e, the total exposure time. This forms in general a (2, 1) exponential family. Note, however, that if we observed until either one of n and t_e took a preassigned value, then it would reduce to a full, i.e., (1, 1), family.

2.4 Choice of priors for exponential family problems

While in Bayesian theory choice of prior is in principle not an issue of achieving mathematical simplicity, nevertheless there are gains in using reasonably simple and flexible forms. In particular, if the likelihood has the full exponential family form

$$m(y) \exp\{s^T \phi - k(\phi)\}, \tag{2.17}$$

a prior for ϕ proportional to

$$\exp\{s_0^T \phi - a_0 k(\phi)\} \tag{2.18}$$

leads to a posterior proportional to

$$\exp\{(s + s_0)^T \phi - (1 + a_0) k(\phi)\}. \tag{2.19}$$

Such a prior is called *conjugate to the likelihood*, or sometimes *closed under sampling*. The posterior distribution has the same form as the prior with s_0 replaced by $s + s_0$ and a_0 replaced by $1 + a_0$.

Example 2.8. *Binomial distribution (ctd).* This continues the discussion of Example 2.5. If the prior for π is proportional to

$$\pi^{r_0}(1 - \pi)^{n_0 - r_0}, \tag{2.20}$$

i.e., is a beta distribution, then the posterior is another beta distribution corresponding to $r + r_0$ successes in $n + n_0$ trials. Thus both prior and posterior are beta distributions. It may help to think of (r_0, n_0) as fictitious data! If the prior information corresponds to fairly small values of n_0 its effect on the conclusions will be small if the amount of real data is appreciable.

2.5 Simple frequentist discussion

In Bayesian approaches sufficiency arises as a convenient simplification of the likelihood; whatever the prior the posterior is formed from the likelihood and hence depends on the data only via the sufficient statistic.

In frequentist approaches the issue is more complicated. Faced with a new model as the basis for analysis, we look for a *Fisherian reduction*, defined as follows:

- find the likelihood function;
- reduce to a sufficient statistic S of the same dimension as θ;
- find a function of S that has a distribution depending only on ψ;
- invert that distribution to obtain limits for ψ at an arbitrary set of probability levels;
- use the conditional distribution of the data given $S = s$ informally or formally to assess the adequacy of the formulation.

Immediate application is largely confined to regular exponential family models.

While most of our discussion will centre on inference about the parameter of interest, ψ, the complementary role of sufficiency of providing an explicit base for model checking is in principle very important. It recognizes that our formulations are always to some extent provisional, and usually capable to some extent of empirical check; the universe of discussion is not closed. In general there is no specification of what to do if the initial formulation is inadequate but, while that might sometimes be clear in broad outline, it seems both in practice and in principle unwise to expect such a specification to be set out in detail in each application.

The next phase of the analysis is to determine how to use s to answer more focused questions, for example about the parameter of interest ψ. The simplest possibility is that there are no nuisance parameters, just a single parameter ψ of interest, and reduction to a single component s occurs. We then have one observation on a known density $f_S(s; \psi)$ and distribution function $F_S(s; \psi)$. Subject to some monotonicity conditions which, in applications, are typically satisfied, the probability statement

$$P(S \leq a_c(\psi)) = F_S(a_c(\psi); \psi) = 1 - c \qquad (2.21)$$

can be inverted for continuous random variables into

$$P\{\psi \leq b_c(S)\} = 1 - c. \qquad (2.22)$$

Thus the statement on the basis of data y, yielding sufficient statistic s, that

$$\psi \leq b_c(s) \qquad (2.23)$$

provides an upper bound for ψ, that is a single member of a hypothetical long run of statements a proportion $1 - c$ of which are true, generating a set of statements in principle at all values of c in $(0, 1)$.

This extends the discussion in Section 1.5, where S is the mean $\bar{Y}$ of n independent and identically normally distributed random variables of known variance and the parameter is the unknown mean.

For discrete distributions there is the complication that the distribution function is a step function so that the inversion operation involved is nonunique. This is an embarrassment typically only when very small amounts of data are involved. We return in Section 3.2 to discuss how best to handle this issue.

An alternative avenue, following the important and influential work of J. Neyman and E. S. Pearson, in a sense proceeds in the reverse direction. Optimality criteria are set up for a formulation of a specific issue, expressing some notion of achieving the most sensitive analysis possible with the available data. The optimality problem is then solved (typically by appeal to a reservoir of relevant theorems) and a procedure produced which, in the contexts under discussion, is then found to involve the data only via s. The two avenues to s nearly, but not quite always, lead to the same destination. This second route will not be followed in the present book.

In the Neyman–Pearson theory of tests, the sensitivity of a test is assessed by the notion of *power*, defined as the probability of reaching a preset level of significance considered for various alternative hypotheses. In the approach adopted here the assessment is via the distribution of the random variable P, again considered for various alternatives. The main applications of power are in the comparison of alternative procedures and in preliminary calculation of appropriate sample size in the design of studies. The latter application, which is inevitably very approximate, is almost always better done by considering the width of confidence intervals.

2.6 Pivots

The following definition is often helpful in dealing with individual applications including those with nuisance parameters.

Suppose that ψ is one-dimensional. Suppose that there is a statistic t, in the present context a function of s but possibly of more general form, and a function $p(t, \psi)$ such that for all $\theta \subset \Omega_\theta$ the random variable $p(T, \psi)$ has a fixed and known continuous distribution and that for all t the function $p(t, \psi)$ is strictly increasing in ψ. Then p is called a *pivot* for inference about ψ and the fixed distribution is called the *pivotal distribution*.

We have that for all ψ and all c, $0 < c < 1$, we can find p_c^* such that

$$P\{p(T, \psi) \leq p_c^*\}, \qquad (2.24)$$

implying that

$$P\{\psi \leq q(T, c)\} = 1 - c. \qquad (2.25)$$

We call $q(t, c)$ a $1 - c$ level upper limit for ψ with obvious changes for lower limits and intervals. The collection of limits for all c encapsulates our information about ψ and the associated uncertainty. In many applications it is convenient to summarize this by giving an upper $1 - c$ limit and a lower $1 - c$ limit forming a $1 - 2c$ *equi-tailed confidence interval*, usually specifying this for one or two conventionally chosen values of c. The use of equal tails is essentially a presentational simplification. Clearly other than equal tails could be used if there were good reason for doing so.

In the special discussion of Section 1.5 the pivot is $(\mu - \bar{Y})/(\sigma_0/\sqrt{n})$ and the pivotal distribution is the standard normal distribution.

Note that the frequentist limits of Section 1.5.2 correspond to the upper c point of the posterior distribution of μ on taking the limit in (1.14) and (1.15) as nv/σ_0^2 increases. Later, in Examples 5.5 and 5.6, the dangers of apparently innocuous priors will be discussed.

The use of pivots is primarily a presentational device. A key notion in the frequentist approach is that of sets of confidence intervals at various levels and the corresponding notion in Bayesian discussions is that of a posterior distribution. At least in relatively simple cases, these are most compactly summarized by a pivot and its distribution. In particular, it may be possible to obtain statistics m_ψ and s_ψ such that in the frequentist approach $(\psi - m_\psi)/s_\psi$ has exactly, or to an adequate approximation, a standard normal distribution, or possibly a Student t distribution. Then confidence intervals at all levels are easily recovered from just the two summarizing statistics. Note that from this perspective these statistics are just a convenient convention for summarization; in particular, it is best not to regard m_ψ as a point estimate of ψ to be considered in isolation.

Example 1.1 can be generalized in many ways. If the standard deviation is unknown and σ_0 is replaced by its standard estimate derived from the residual sum of squares, the pivotal distribution becomes the Student t distribution with $n - 1$ degrees of freedom. Further if we consider other than sufficient statistics, the mean could be replaced by the median or by any other location estimate with a corresponding modification of the pivotal distribution. The argument applies directly to the estimation of any linear parameter in the linear model of Example 1.4. In the second-order linear model generalizing Example 1.3, the pivotal distribution is asymptotically standard normal by the Central Limit Theorem and, if the variance is estimated, also by appeal to the Weak Law of Large Numbers.

In the previous examples the parameter of interest has been a scalar. The ideas extend to vector parameters of interest ψ, although in most applications it is often preferable to work with single interpretable components, often representing contrasts or measures of dependence interpreted one at a time.

Example 2.9. *Mean of a multivariate normal distribution.* The most important example of a vector parameter of interest concerns the vector mean μ of a multivariate normal distribution. Suppose that $Y_1, \ldots, Y_n$ are independent and identically distributed $p \times 1$ vectors having a multivariate normal distribution of mean μ and known nonsingular covariance matrix Σ_0.

Evaluation of the likelihood shows that the sample mean is a minimal sufficient statistic and that it has the same dimension as the unknown parameter. It can be shown by linear transformation to a set of independent random components that the quadratic form $n(\bar{Y}-\mu)^T \Sigma_0^{-1}(\bar{Y}-\mu)$ has a chi-squared distribution with p degrees of freedom and so forms a pivot. There is thus a set of concentric similar ellipsoids defined by Σ_0 such that $\bar{Y}$ lies within each with specified probability. Inversion of this yields confidence ellipsoids for μ centred on $\bar{y}$.

These regions are *likelihood-based* in the sense that any value of μ excluded from a confidence region has lower likelihood than all parameter values μ included in that region. This distinguishes these regions from those of other shapes, for example from rectangular boxes with edges parallel to the coordinate axes. The elliptical regions are also invariant under arbitrary nonsingular linear transformation of the vectors Y.

Quadratic forms associated in this way with the inverse of a covariance matrix occur frequently, especially in Chapter 6, and we note their geometrical interpretation. See Note 2.6.

Notes 2

Section 2.1. In some contexts the likelihood is better defined as the equivalence class of all functions obtained by multiplying by an arbitrary positive function of y. This would formalize the notion that it is really only the ratio of the likelihoods at two different parameter values that is inferentially meaningful.

The importance of likelihood was emphasized quite often by R. A. Fisher, who also suggested that for some types of application it alone was appropriate; see Fisher (1956). For accounts of statistics putting primary emphasis on likelihood as such, see Edwards (1992) and Royall (1997) and, in a particular time series context, Barnard et al. (1962). The expression *law of the likelihood* was introduced by Hacking (1965). In Bayesian theory the likelihood captures the information directly in the data.

Section 2.2. The Jacobian is the matrix of partial derivatives involved in any change of variables in a multiple integral and so, in particular, in any change of variables in a probability density; see any textbook on calculus.

Sufficiency was introduced by Fisher (1922). A very mathematical treatment is due to Halmos and Savage (1949). The property arises in all approaches to statistical inference although its conceptual importance, heavily emphasized in the present book, depends greatly on the viewpoint taken.

Section 2.3. The exponential family was first studied systematically in the 1930s by G. Darmois, B. O. Koopmans and E. J. G. Pitman after some combination of whom it is sometimes named. Barndorff-Nielsen (1978) gives a systematic account stressing the links with convex analysis. Essentially all well-behaved models with sufficient statistic having the same dimension as the parameter space are of this simple exponential family form. The general books listed in Notes 1 all give some account of the exponential family.

The cumulant (or semi-invariant) generating function is the log of the moment generating function, i.e., $\log E(e^{pY})$. If it can be expanded in a power series $\Sigma \kappa_r p^r / r!$, the κ_r are cumulants. The first cumulant is the mean, the higher cumulants are functions of moments about the mean with $\kappa_2 = \text{var}(Y)$. For a normal distribution the cumulants of order 3 and higher are zero and for the multivariate version $\kappa_{11} = \text{cov}(Y_1, Y_2)$. The cumulant generating function of the sum of independent random variables is the sum of the separate cumulant generating functions with the corresponding property holding also for the cumulants.

The notation ∇ is used for the gradient operator, forming a column vector of partial derivatives. It is intended here and especially in Chapter 6 to show multiparameter arguments as closely parallel to the corresponding single-parameter versions. It is necessary only to see that $\nabla \nabla^T$ forms a square matrix of second partial derivatives and later that Taylor's theorem up to second-degree terms can be written

$$g(x+h) = g(x) + \nabla^T g(x) h + h^T \nabla \nabla^T g(x) h / 2,$$

where g is a real-valued function of the multidimensional argument x.

Section 2.5. Definitive accounts of the Neyman–Pearson theory are given in two books by E. L. Lehmann which have acquired coauthors as successive revisions have been produced (Lehmann and Casella, 2001; Lehmann and Romano, 2004). The original papers are available in collected form (Neyman and Pearson, 1967).

Section 2.6. The notion of a pivot was given a central role in influential but largely unpublished work by G. A. Barnard. The basis of Example 2.9 and its further development is set out in books on multivariate analysis; see especially Anderson (2004). The starting point is that the vector mean $\bar{Y}$ has a multivariate normal distribution of mean μ and covariance matrix Σ_0/n, a result proved by generalizing the argument in Note 1.5.

The Weak Law of Large Numbers is essentially that the average of a large number of independent (or not too strongly dependent) random variables is, under rather general conditions, with high probability close to the average of the expected values. It uses a particular definition of convergence, convergence in probability. The Strong Law, which forms a pinnacle of classical probability theory, uses a different notion of convergence which is, however, not relevant in the statistical problems treated here. See also Note 6.2.

For the result of Example 2.9 and for many other developments the following is needed. If Z is a $p \times 1$ vector of zero mean its covariance matrix is $\Sigma_Z = E(ZZ^T)$. If c is an arbitrary $p \times p$ nonsingular matrix, $\Sigma_{cZ} = c\Sigma_Z c^T$ and $\Sigma_{cZ}^{-1} = (c^T)^{-1}\Sigma_Z^{-1} c^{-1}$. A direct calculation now shows that $Z^T \Sigma_Z^{-1} Z = (cZ)^T \Sigma_{cZ}^{-1}(cZ)$, so that if we define $\|Z\|_{\Sigma_Z}^2 = Z^T \Sigma_Z^{-1} Z$ its value is invariant under nonsingular linear transformations. If, in particular, we choose c so that $\Sigma_{cZ} = I$, the identity matrix, the quadratic form reduces to the sum of squares of the components, i.e., the Euclidean squared length of the vector. If the components are normally distributed the distribution of the sum of squares is the chi-squared distribution with p degrees of freedom.

We may thus reasonably call $\|Z\|_{\Sigma_Z}$ the norm of the vector Z in the metric defined by its covariance matrix. It is in a sense the natural measure of length attached to the ellipsoid defined by the covariance matrix.

3
Significance tests

Summary. First a number of distinct situations are given in which significance tests may be relevant. The nature of a simple significance test is set out and its implications explored. The relation with interval estimation is emphasized. While most of the discussion is from a frequentist perspective, relations with Bayesian theory are outlined in the final section.

3.1 General remarks

So far, in our frequentist discussion we have summarized information about the unknown parameter ψ by finding procedures that would give in hypothetical repeated applications upper (or lower) bounds for ψ a specified proportion of times in a long run of repeated applications. This is close to but not the same as specifying a probability distribution for ψ; it avoids having to treat ψ as a random variable, and moreover as one with a *known* distribution in the absence of the data.

Suppose now there is specified a particular value ψ_0 of the parameter of interest and we wish to assess the relation of the data to that value. Often the hypothesis that $\psi = \psi_0$ is called the *null hypothesis* and conventionally denoted by H_0. It may, for example, assert that some effect is zero or takes on a value given by a theory or by previous studies, although ψ_0 does not have to be restricted in that way.

There are at least six different situations in which this may arise, namely the following.

- There may be some special reason for thinking that the null hypothesis may be exactly or approximately true or strong subject-matter interest may focus on establishing that it is likely to be false.

- There may be no special reason for thinking that the null hypothesis is true but it is important because it divides the parameter space into two (or more) regions with very different interpretations. We are then interested in whether the data establish reasonably clearly which region is correct, for example it may establish the value of $\text{sgn}(\psi - \psi_0)$.
- Testing may be a technical device used in the process of generating confidence intervals.
- Consistency with $\psi = \psi_0$ may provide a reasoned justification for simplifying an otherwise rather complicated model into one that is more transparent and which, initially at least, may be a reasonable basis for interpretation.
- Only the model when $\psi = \psi_0$ is under consideration as a possible model for interpreting the data and it has been embedded in a richer family just to provide a qualitative basis for assessing departure from the model.
- Only a single model is defined, but there is a qualitative idea of the kinds of departure that are of potential subject-matter interest.

The last two formulations are appropriate in particular for examining model adequacy.

From time to time in the discussion it is useful to use the short-hand description of H_0 as being possibly *true*. Now in statistical terms H_0 refers to a probability model and the very word 'model' implies idealization. With a very few possible exceptions it would be absurd to think that a mathematical model is an exact representation of a real system and in that sense all H_0 are defined within a system which is untrue. We use the term to mean that in the current state of knowledge it is reasonable to proceed as if the hypothesis is true. Note that an underlying subject-matter hypothesis such as that a certain environmental exposure has absolutely no effect on a particular disease outcome might indeed be true.

3.2 Simple significance test

In the formulation of a simple significance test, we suppose available data y and a null hypothesis H_0 that specifies the distribution of the corresponding random variable Y. In the first place, no other probabilistic specification is involved, although some notion of the type of departure from H_0 that is of subject-matter concern is essential.

The first step in testing H_0 is to find a distribution for observed random variables that has a form which, under H_0, is free of nuisance parameters, i.e., is completely known. This is trivial when there is a single unknown parameter whose value is precisely specified by the null hypothesis. Next find or determine a test statistic T, large (or extreme) values of which indicate a departure from the null hypothesis of subject-matter interest. Then if t_{obs} is the observed value of T we define

$$p_{\text{obs}} = P(T \geq t_{\text{obs}}), \tag{3.1}$$

the probability being evaluated under H_0, to be the (observed) *p-value* of the test.

It is conventional in many fields to report only very approximate values of p_{obs}, for example that the departure from H_0 is significant just past the 1 per cent level, etc.

The hypothetical frequency interpretation of such reported significance levels is as follows. If we were to accept the available data as just decisive evidence against H_0, then we would reject the hypothesis when true a long-run proportion p_{obs} of times.

Put more qualitatively, we examine consistency with H_0 by finding the consequences of H_0, in this case a random variable with a known distribution, and seeing whether the prediction about its observed value is reasonably well fulfilled.

We deal first with a very special case involving testing a null hypothesis that might be true to a close approximation.

Example 3.1. *Test of a Poisson mean.* Suppose that Y has a Poisson distribution of unknown mean μ and that it is required to test the null hypothesis $\mu = \mu_0$, where μ_0 is a value specified either by theory or a large amount of previous experience. Suppose also that only departures in the direction of larger values of μ are of interest. Here there is no ambiguity about the choice of test statistic; it has to be Y or a monotone function of Y and given that $Y = y$ the *p*-value is

$$p_{\text{obs}} = \Sigma_{v=y}^{\infty} e^{-\mu_0} \mu_0^v / v!. \tag{3.2}$$

Now suppose that instead of a single observation we have n independent replicate observations so that the model is that $Y_1, \ldots, Y_n$ have independent Poisson distributions all with mean μ. With the same null hypothesis as before, there are now many possible test statistics that might be used, for example $\max(Y_k)$. A preference for sufficient statistics leads, however, to the use of ΣY_k, which under the null hypothesis has a Poisson distribution of mean $n\mu_0$. Then the *p*-value is again given by (3.2), now with μ_0 replaced by $n\mu_0$ and with y replaced by the observed value of ΣY_k.

We return to this illustration in Example 3.3. The testing of a null hypothesis about the mean μ of a normal distribution when the standard deviation σ_0 is known follows the same route. The distribution of the sample mean $\bar{Y}$ under the null hypothesis $\mu = \mu_0$ is now given by an integral rather than by a sum and (3.2) is replaced by

$$p_{\text{obs}} = 1 - \Phi\left(\frac{\bar{y} - \mu_0}{\sigma_0/\sqrt{n}}\right). \tag{3.3}$$

We now turn to a complementary use of these ideas, namely to test the adequacy of a given model, what is also sometimes called model criticism. We illustrate this by testing the adequacy of the Poisson model. It is necessary if we are to parallel the previous argument to find a statistic whose distribution is exactly or very nearly independent of the unknown parameter μ. An important way of doing this is by appeal to the second property of sufficient statistics, namely that after conditioning on their observed value the remaining data have a fixed distribution.

Example 3.2. *Adequacy of Poisson model.* Let $Y_1, \ldots, Y_n$ be independent Poisson variables with unknown mean μ. The null hypothesis H_0 for testing model adequacy is that this model applies for some unknown μ. Initially no alternative is explicitly formulated. The sufficient statistic is ΣY_k, so that to assess consistency with the model we examine the conditional distribution of the data given $\Sigma Y_k = s$. This density is zero if $\Sigma y_k \neq s$ and is otherwise

$$\frac{s!}{\Pi y_k!} \frac{1}{n^s}, \tag{3.4}$$

i.e., is a multinomial distribution with s trials each giving a response equally likely to fall in one of n cells. Because this distribution is completely specified numerically, we are essentially in the same situation as if testing consistency with a null hypothesis that completely specified the distribution of the observations free of unknown parameters. There remains, except when $n = 2$, the need to choose a test statistic. This is usually taken to be either the *dispersion index* $\Sigma(Y_k - \bar{Y})^2/\bar{Y}$ or the number of zeros.

The former is equivalent to the ratio of the sample estimate of variance to the mean. In this conditional specification, because the sample total is fixed, the statistic is equivalent also to ΣY_k^2. Note that if, for example, the dispersion test is used, no explicit family of alternative models has been specified, only an indication of the kind of discrepancy that it is especially important to detect. A more formal and fully parametric procedure might have considered the negative binomial distribution as representing such departures and then used the apparatus of the Neyman–Pearson theory of testing hypotheses to develop a test especially sensitive to such departures.

A quite high proportion of the more elementary tests used in applications were developed by the relatively informal route just outlined. When a full family of distributions is specified, covering both the null hypothesis and one or more alternatives representing important departures from H_0, it is natural to base the test on the optimal statistic for inference about ψ within that family. This typically has sensitivity properties in making the random variable P corresponding to p_{obs} stochastically small under alternative hypotheses.

For continuously distributed test statistics, p_{obs} typically may take any real value in $(0, 1)$. In the discrete case, however, only a discrete set of values of p are achievable in any particular case. Because preassigned values such as 0.05 play no special role, the only difficulty in interpretation is the theoretical one of comparing alternative procedures with different sets of achievable values.

Example 3.3. *More on the Poisson distribution.* For continuous observations the random variable P can, as already noted, in principle take any value in $(0, 1)$. Suppose, however, we return to the special case that Y has a Poisson distribution with mean μ and that the null hypothesis $\mu = \mu_0$ is to be tested checking for departures in which $\mu > \mu_0$, or more generally in which the observed random variable Y is stochastically larger than a Poisson-distributed random variable of mean μ_0. Then for a given observation y the p-value is

$$p_{\text{obs}}^{+} = \Sigma_{v=y}^{\infty} e^{-\mu_0} \mu_0^v / v!, \tag{3.5}$$

whereas for detecting departures in the direction of small values the corresponding p-value is

$$p_{\text{obs}}^{-} = \Sigma_{v=0}^{y} e^{-\mu_0} \mu_0^v / v!. \tag{3.6}$$

Table 3.1 shows some values for the special case $\mu_0 = 2$. So far as use for a one-sided significance test is concerned, the restriction to a particular set of values is unimportant unless that set is in some sense embarrassingly small. Thus the conclusion from Table 3.1(b) that in testing $\mu = 2$ looking for departures $\mu < 2$ even the most extreme observation possible, namely zero, does not have a particularly small p-value is hardly surprising. We return to the implications for two-sided testing in the next section.

In forming upper confidence limits for μ based on an observed y there is no difficulty in finding critical values of μ such that the relevant lower tail area is some assigned value. Thus with $y = 0$ the upper 0.95 point for μ is such that $e^{-\mu^*} = 0.05$, i.e., $\mu^* = \log 20 \simeq 3$. A similar calculation for a lower confidence limit is not possible for $y = 0$, but is possible for all other y. The discreteness of the set of achievable p-values is in this context largely unimportant.

Table 3.1. *Achievable significance levels for testing that a Poisson-distributed random variable with observed value y has mean 2: (a) test against alternatives larger than 2; (b) test against alternatives less than 2*

(a)	y	2	3	4	5	6
	p	0.594	0.323	0.143	0.053	0.017
(b)	y	0	1	2		
	p	0.135	0.406	0.677		

3.3 One- and two-sided tests

In many situations observed values of the test statistic in either tail of its distribution represent interpretable, although typically different, departures from H_0. The simplest procedure is then often to contemplate two tests, one for each tail, in effect taking the more significant, i.e., the smaller tail, as the basis for possible interpretation. Operational interpretation of the result as a hypothetical error rate is achieved by doubling the corresponding p, with a slightly more complicated argument in the discrete case.

More explicitly we argue as follows. With test statistic T, consider two p-values, namely

$$p_{\text{obs}}^+ = P(T \geq t; H_0), \quad p_{\text{obs}}^- = P(T \leq t; H_0). \tag{3.7}$$

In general the sum of these values is $1 + P(T = t)$. In the two-sided case it is then reasonable to define a new test statistic

$$Q = \min(P_{\text{obs}}^+, P_{\text{obs}}^-). \tag{3.8}$$

The level of significance is

$$P(Q \leq q_{\text{obs}}; H_0). \tag{3.9}$$

In the continuous case this is $2q_{\text{obs}}$ because two disjoint events are involved. In a discrete problem it is q_{obs} plus the achievable p-value from the other tail of the distribution nearest to but not exceeding q_{obs}. As has been stressed the precise calculation of levels of significance is rarely if ever critical, so that the careful definition is more one of principle than of pressing applied importance. A more important point is that the definition is unaffected by a monotone transformation of T.

In one sense very many applications of tests are essentially two-sided in that, even though initial interest may be in departures in one direction, it will rarely be wise to disregard totally departures in the other direction, even if initially they are unexpected. The interpretation of differences in the two directions may well be very different. Thus in the broad class of procedures associated with the linear model of Example 1.4 tests are sometimes based on the ratio of an estimated variance, expected to be large if real systematic effects are present, to an estimate essentially of error. A large ratio indicates the presence of systematic effects whereas a suspiciously small ratio suggests an inadequately specified model structure.

3.4 Relation with acceptance and rejection

There is a conceptual difference, but essentially no mathematical difference, between the discussion here and the treatment of testing as a two-decision problem, with control over the formal error probabilities. In this we fix in principle the probability of rejecting H_0 when it is true, usually denoted by α, aiming to maximize the probability of rejecting H_0 when false. This approach demands the explicit formulation of alternative possibilities. Essentially it amounts to setting in advance a threshold for p_{obs}. It is, of course, potentially appropriate when clear decisions are to be made, as for example in some classification problems. The previous discussion seems to match more closely scientific practice in these matters, at least for those situations where analysis and interpretation rather than decision-making are the focus.

That is, there is a distinction between the Neyman–Pearson formulation of testing regarded as clarifying the meaning of statistical significance via hypothetical repetitions and that same theory regarded as in effect an instruction on how to implement the ideas by choosing a suitable α in advance and reaching different decisions accordingly. The interpretation to be attached to accepting or rejecting a hypothesis is strongly context-dependent; the point at stake here, however, is more a question of the distinction between assessing evidence, as contrasted with deciding by a formal rule which of two directions to take.

3.5 Formulation of alternatives and test statistics

As set out above, the simplest version of a significance test involves formulation of a null hypothesis H_0 and a test statistic T, large, or possibly extreme, values of which point against H_0. Choice of T is crucial in specifying the kinds of departure from H_0 of concern. In this first formulation no alternative probability models are explicitly formulated; an implicit family of possibilities is

3.5 Formulation of alternatives and test statistics

specified via T. In fact many quite widely used statistical tests were developed in this way.

A second possibility is that the null hypothesis corresponds to a particular parameter value, say $\psi = \psi_0$, in a family of models and the departures of main interest correspond either, in the one-dimensional case, to one-sided alternatives $\psi > \psi_0$ or, more generally, to alternatives $\psi \neq \psi_0$. This formulation will suggest the most sensitive test statistic, essentially equivalent to the best estimate of ψ, and in the Neyman–Pearson formulation such an explicit formulation of alternatives is essential.

The approaches are, however, not quite as disparate as they may seem. Let $f_0(y)$ denote the density of the observations under H_0. Then we may associate with a proposed test statistic T the exponential family

$$f_0(y)\exp\{t\theta - k(\theta)\}, \qquad (3.10)$$

where $k(\theta)$ is a normalizing constant. Then the test of $\theta = 0$ most sensitive to these departures is based on T. Not all useful tests appear natural when viewed in this way, however; see, for instance, Example 3.5.

Many of the test procedures for examining model adequacy that are provided in standard software are best regarded as defined directly by the test statistic used rather than by a family of alternatives. In principle, as emphasized above, the null hypothesis is the conditional distribution of the data given the sufficient statistic for the parameters in the model. Then, within that null distribution, interesting directions of departure are identified.

The important distinction is between situations in which a whole family of distributions arises naturally as a base for analysis versus those where analysis is at a stage where only the null hypothesis is of explicit interest.

Tests where the null hypotheses itself is formulated in terms of arbitrary distributions, so-called *nonparametric* or *distribution-free* tests, illustrate the use of test statistics that are formulated largely or wholly informally, without specific probabilistically formulated alternatives in mind. To illustrate the arguments involved, consider initially a single homogenous set of observations.

That is, let $Y_1, \ldots, Y_n$ be independent and identically distributed random variables with arbitrary cumulative distribution function $F(y) = P(Y_k \leq y)$. To avoid minor complications we suppose throughout the following discussion that the distribution is continuous so that, in particular, the possibility of ties, i.e., exactly equal pairs of observations, can be ignored. A formal likelihood can be obtained by dividing the real line into a very large number of very small intervals each having an arbitrary probability attached to it. The likelihood for data y can then be specified in a number of ways, namely by

- a list of data values actually observed,
- the *sample cumulative distribution function*, defined as

$$F_n(y) = n^{-1} \Sigma I(y_k \leq y), \qquad (3.11)$$

where the indicator function $I(y_k \leq y)$ is one if $y_k \leq y$ and zero otherwise,
- the set of order statistics $y_{(1)} \leq y_{(2)} \leq \cdots \leq y_{(n)}$, i.e., the observed values arranged in increasing order.

The second and third of these are reductions of the first, suppressing the information about the order in which the observations are obtained. In general no further reduction is possible and so either of the last two forms the sufficient statistic. Thus if we apply the general prescription, conclusions about the function $F(y)$ are to be based on one of the above, for example on the sample distribution function, whereas consistency with the model is examined via the conditional distribution given the order statistics. This conditional distribution specifies that starting from the original data, all $n!$ permutations of the data are equally likely. This leads to tests in general termed *permutation tests*. This idea is now applied to slightly more complicated situations.

Example 3.4. *Test of symmetry.* Suppose that the null hypothesis is that the distribution is symmetric about a known point, which we may take to be zero. That is, under the null hypothesis $F(-y) + F(y) = 1$. Under this hypothesis, all points y and $-y$ have equal probability, so that the sufficient statistic is determined by the order statistics or sample distribution function of the $|y_k|$. Further, conditionally on the sufficient statistic, all 2^n sample points $\pm y_k$ have equal probability 2^{-n}. Thus the distribution under the null hypothesis of any test statistic is in principle exactly known.

Simple one-sided test statistics for symmetry can be based on the number of positive observations, leading to the *sign test*, whose null distribution is binomial with parameter $\frac{1}{2}$ and index n, or on the mean of all observations. The distribution of the latter can be found by enumeration or approximated by finding its first few moments and using a normal or other approximation to the distribution.

This formulation is relevant, for example, when the observations analyzed are differences between primary observations after and before some intervention, or are differences obtained from the same individual under two different regimes.

A strength of such procedures is that they involve no specification of the functional form of $F(y)$. They do, however, involve strong independence assumptions, which may often be more critical. Moreover, they do not extend easily to relatively complicated models.

3.5 Formulation of alternatives and test statistics

Example 3.5. *Nonparametric two-sample test.* Let $(Y_{11}, \ldots, Y_{1n_1}; Y_{21}, \ldots, Y_{2n_2})$ be two sets of mutually independent random variables with cumulative distribution functions respectively $F_1(y)$ and $F_2(y)$. Consider the null hypothesis $F_1(y) = F_2(y)$ for all y.

When this hypothesis is true the sufficient statistic is the set of order statistics of the combined set of observations and under this hypothesis all $(n_1 + n_2)!$ permutations of the data are equally likely and, in particular, the first set of n_1 observations is in effect a random sample drawn without replacement from the full set, allowing the null distribution of any test statistic to be found.

Sometimes it may be considered that while the ordering of possible observational points is meaningful the labelling by numerical values is not. Then we look for procedures invariant under arbitrary strictly monotone increasing transformations of the measurement scale. This is achieved by replacing the individual observations y by their rank order in the full set of order statistics of the combined sample. If the test statistic, T_W, is the sum of the ranks of, say, the first sample, the resulting test is called the *Wilcoxon rank sum test* and the parallel test for the single-sample symmetry problem is the *Wilcoxon signed rank test*.

The distribution of the test statistic under the null hypothesis, and hence the level of significance, is in principle found by enumeration. The moments of T_W under the null hypothesis can be found by the arguments to be developed in a rather different context in Section 9.2 and in fact the mean and variance are respectively $n_1(n_1 + n_2 + 1)/2$ and $n_1 n_2(n_1 + n_2 + 1)/12$. A normal approximation, with continuity correction, based on this mean and variance will often be adequate.

Throughout this discussion the full set of values of y is regarded as fixed.

The choice of test statistic in these arguments is based on informal considerations or broad analogy. Sometimes, however, the choice can be sharpened by requiring good sensitivity of the test were the data produced by some unknown monotonic transformation of data following a specific parametric model.

For the two-sample problem, the most obvious possibilities are that the data are transformed from underlying normal or underlying exponential distributions, a test of equality of the relevant means being required in each case. The exponential model is potentially relevant for the analysis of survival data. Up to a scale and location change in the normal case and up to a scale change in the exponential case, the originating data can then be reconstructed approximately under the null hypothesis by replacing the rth largest order statistic out of n in the data by the expected value of that order statistic in samples from the standard normal or unit exponential distribution respectively. Then the standard parametric test statistics are used, relying in principle on their permutation distribution to preserve the nonparametric propriety of the test.

It can be shown that, purely as a test procedure, the loss of sensitivity of the resulting nonparametric analysis as compared with the fully parametric analysis, were it available, is usually small. In the normal case the expected order statistics, called Fisher and Yates scores, are tabulated, or can be approximated by $\Phi^{-1}\{(r - 3/8)/(n + 1/4)\}$. For the exponential distribution the scores can be given in explicit form or approximated by $\log\{(n + 1)/(n + 1 - r - 1/2)\}$.

3.6 Relation with interval estimation

While conceptually it may seem simplest to regard estimation with uncertainty as a simpler and more direct mode of analysis than significance testing there are some important advantages, especially in dealing with relatively complicated problems, in arguing in the other direction. Essentially confidence intervals, or more generally confidence sets, can be produced by testing consistency with every possible value in Ω_ψ and taking all those values not 'rejected' at level c, say, to produce a $1 - c$ level interval or region. This procedure has the property that in repeated applications any true value of ψ will be included in the region except in a proportion $1 - c$ of cases. This can be done at various levels c, using the same form of test throughout.

Example 3.6. *Ratio of normal means.* Given two independent sets of random variables from normal distributions of unknown means μ_0, μ_1 and with known variance σ_0^2, we first reduce by sufficiency to the sample means $\bar{y}_0, \bar{y}_1$. Suppose that the parameter of interest is $\psi = \mu_1/\mu_0$. Consider the null hypothesis $\psi = \psi_0$. Then we look for a statistic with a distribution under the null hypothesis that does not depend on the nuisance parameter. Such a statistic is

$$\frac{\bar{Y}_1 - \psi_0 \bar{Y}_0}{\sigma_0 \sqrt{(1/n_1 + \psi_0^2/n_0)}}; \tag{3.12}$$

this has a standard normal distribution under the null hypothesis. This with ψ_0 replaced by ψ could be treated as a pivot provided that we can treat $\bar{Y}_0$ as positive.

Note that provided the two distributions have the same variance a similar result with the Student t distribution replacing the standard normal would apply if the variance were unknown and had to be estimated. To treat the probably more realistic situation where the two distributions have different and unknown variances requires the approximate techniques of Chapter 6.

We now form a $1-c$ level confidence region by taking all those values of ψ_0 that would not be 'rejected' at level c in this test. That is, we take the set

$$\left\{ \psi : \frac{(\bar{Y}_1 - \psi \bar{Y}_0)^2}{\sigma_0^2(1/n_1 + \psi^2/n_0)} \leq k^*_{1;\,c} \right\}, \qquad (3.13)$$

where $k^*_{1;\,c}$ is the upper c point of the chi-squared distribution with one degree of freedom.

Thus we find the limits for ψ as the roots of a quadratic equation. If there are no real roots, *all* values of ψ are consistent with the data at the level in question. If the numerator and especially the denominator are poorly determined, a confidence interval consisting of the whole line may be the only rational conclusion to be drawn and is entirely reasonable from a testing point of view, even though regarded from a confidence interval perspective it may, wrongly, seem like a vacuous statement.

Depending on the context, emphasis may lie on the possible explanations of the data that are reasonably consistent with the data or on those possible explanations that have been reasonably firmly refuted.

Example 3.7. *Poisson-distributed signal with additive noise.* Suppose that Y has a Poisson distribution with mean $\mu + a$, where $a > 0$ is a known constant representing a background process of noise whose rate of occurrence has been estimated with high precision in a separate study. The parameter μ corresponds to a signal of interest. Now if, for example, $y = 0$ and a is appreciable, for example, $a \geq 4$, when we test consistency with each possible value of μ *all* values of the parameter are inconsistent with the data until we use very small values of c. For example, the 95 per cent confidence interval will be empty. Now in terms of the initial formulation of confidence intervals, in which, in particular, the model is taken as a firm basis for analysis, this amounts to making a statement that is certainly wrong; there is by supposition some value of the parameter that generated the data. On the other hand, regarded as a statement of which values of μ are consistent with the data at such-and-such a level the statement is perfectly reasonable and indeed is arguably the only sensible frequentist conclusion possible at that level of c.

3.7 Interpretation of significance tests

There is a large and ever-increasing literature on the use and misuse of significance tests. This centres on such points as:

1. Often the null hypothesis is almost certainly false, inviting the question why is it worth testing it?

2. Estimation of ψ is usually more enlightening than testing hypotheses about ψ.
3. Failure to 'reject' H_0 does not mean that we necessarily consider H_0 to be exactly or even nearly true.
4. If tests show that data are consistent with H_0 and inconsistent with the minimal departures from H_0 considered as of subject-matter importance, then this may be taken as positive support for H_0, i.e., as more than mere consistency with H_0.
5. With large amounts of data small departures from H_0 of no subject-matter importance may be highly significant.
6. When there are several or many somewhat similar sets of data from different sources bearing on the same issue, separate significance tests for each data source on its own are usually best avoided. They address individual questions in isolation and this is often inappropriate.
7. p_{obs} is not the probability that H_0 is true.

Discussion of these points would take us too far afield. Point 7 addresses a clear misconception. The other points are largely concerned with how in applications such tests are most fruitfully applied and with the close connection between tests and interval estimation. The latter theme is emphasized below. The essential point is that significance tests in the first place address the question of whether the data are reasonably consistent with a null hypothesis in the respect tested. This is in many contexts an interesting but limited question. The much fuller specification needed to develop confidence limits by this route leads to much more informative summaries of what the data plus model assumptions imply.

3.8 Bayesian testing

A Bayesian discussion of significance testing is available only when a full family of models is available. We work with the posterior distribution of ψ. When the null hypothesis is quite possibly exactly or nearly correct we specify a prior probability π_0 that H_0 is true; we need also to specify the conditional prior distribution of ψ when H_0 is false, as well as aspects of the prior distribution concerning nuisance parameters λ. Some care is needed here because the issue of testing is not likely to arise when massive easily detected differences are present. Thus when, say, ψ can be estimated with a standard error of $\sigma_0/\sqrt{n}$

the conditional prior should have standard deviation $b\sigma_0/\sqrt{n}$, for some not too large value of b.

When the role of H_0 is to divide the parameter space into qualitatively different parts the discussion essentially is equivalent to checking whether the posterior interval at some suitable level overlaps the null hypothesis value of ψ. If and only if there is no overlap the region containing ψ is reasonably firmly established. In simple situations, such as that of Example 1.1, posterior and confidence intervals are in exact or approximate agreement when flat priors are used, providing in such problems some formal justification for the use of flat priors or, from a different perspective, for confidence intervals.

We defer to Section 5.12 the general principles that apply to the choice of prior distributions, in particular as they affect both types of testing problems mentioned here.

Notes 3

Section 3.1. The explicit classification of types of null hypothesis is developed from Cox (1977).

Section 3.2. The use of the conditional distribution to test conformity with a Poisson distribution follows Fisher (1950). Another route for dealing with discrete distributions is to define (Stone, 1969) the p-value for test statistic T by $P(T > t_{\text{obs}}) + P(T = t_{\text{obs}})/2$. This produces a statistic having more nearly a uniform distribution under the null hypothesis but the motivating operational meaning has been sacrificed.

Section 3.3. There are a number of ways of defining two-sided p-values for discrete distributions; see, for example, Cox and Hinkley (1974, p.79).

Section 3.4. The contrast made here between the calculation of p-values as measures of evidence of consistency and the more decision-focused emphasis on accepting and rejecting hypotheses might be taken as one characteristic difference between the Fisherian and the Neyman–Pearson formulations of statistical theory. While this is in some respects the case, the actual practice in specific applications as between Fisher and Neyman was almost the reverse. Neyman often in effect reported p-values whereas some of Fisher's use of tests in applications was much more dichotomous. For a discussion of the notion of severity of tests, and the circumstances when consistency with H_0 might be taken as positive support for H_0, see Mayo (1996).

Section 3.5. For a thorough account of nonparametric tests, see Lehmann (1998).

Section 3.6. The argument for the ratio of normal means is due to E. C. Fieller, after whom it is commonly named. The result applies immediately to the ratio of least squares regression coefficients and hence in particular to estimating the intercept of a regression line on the z-coordinate axis. A substantial dispute developed over how this problem should be handled from the point of view of Fisher's fiducial theory, which mathematically but not conceptually amounts to putting flat priors on the two means (Creasy, 1954). There are important extensions of the situation, for example to inverse regression or (controlled) calibration, where on the basis of a fitted regression equation it is desired to estimate the value of an explanatory variable that generated a new value of the response variable.

Section 3.8. For more on Bayesian tests, see Jeffreys (1961) and also Section 6.2.6 and Notes 6.2.

4

More complicated situations

Summary. This chapter continues the comparative discussion of frequentist and Bayesian arguments by examining rather more complicated situations. In particular several versions of the two-by-two contingency table are compared and further developments indicated. More complicated Bayesian problems are discussed.

4.1 General remarks

The previous frequentist discussion in especially Chapter 3 yields a theoretical approach which is limited in two senses. It is restricted to problems with no nuisance parameters or ones in which elimination of nuisance parameters is straightforward. An important step in generalizing the discussion is to extend the notion of a Fisherian reduction. Then we turn to a more systematic discussion of the role of nuisance parameters.

By comparison, as noted previously in Section 1.5, a great formal advantage of the Bayesian formulation is that, once the formulation is accepted, all subsequent problems are computational and the simplifications consequent on sufficiency serve only to ease calculations.

4.2 General Bayesian formulation

The argument outlined in Section 1.5 for inference about the mean of a normal distribution can be generalized as follows. Consider the model $f_{Y|\Theta}(y \mid \theta)$, where, because we are going to treat the unknown parameter as a random variable, we now regard the model for the data-generating process as a conditional density. Suppose that Θ has the prior density $f_\Theta(\theta)$, specifying the marginal distribution of the parameter, i.e., in effect the distribution Θ has when the observations y are not available.

Given the data and the above formulation it is reasonable to assume that all information about θ is contained in the conditional distribution of Θ given $Y = y$. We call this the posterior distribution of Θ and calculate it by the standard laws of probability theory, as given in (1.12), by

$$f_{\Theta|Y}(\theta \mid y) = \frac{f_{Y|\Theta}(y \mid \theta) f_{\Theta}(\theta)}{\int_{\Omega_\theta} f_{Y|\Theta}(y \mid \phi) f_{\Theta}(\phi) d\phi}. \tag{4.1}$$

The main problem in computing this lies in evaluating the normalizing constant in the denominator, especially if the dimension of the parameter space is high.

Finally, to isolate the information about the parameter of interest ψ, we marginalize the posterior distribution over the nuisance parameter λ. That is, writing $\theta = (\psi, \lambda)$, we consider

$$f_{\Psi|Y}(\psi \mid y) = \int f_{\Theta|Y}(\psi, \lambda \mid y) d\lambda. \tag{4.2}$$

The models and parameters for which this leads to simple explicit solutions are broadly those for which frequentist inference yields simple solutions.

Because in the formula for the posterior density the prior density enters both in the numerator and the denominator, formal multiplication of the prior density by a constant would leave the answer unchanged. That is, there is no need for the prior measure to be normalized to 1 so that, formally at least, improper, i.e. divergent, prior densities may be used, always provided proper posterior densities result. A simple example in the context of the normal mean, Section 1.5, is to take as prior density element $f_M(\mu) d\mu$ just $d\mu$. This could be regarded as the limit of the proper normal prior with variance v taken as $v \to \infty$. Such limits raise few problems in simple cases, but in complicated multiparameter problems considerable care would be needed were such limiting notions contemplated. There results here a posterior distribution for the mean that is normal with mean $\bar{y}$ and variance σ_0^2/n, leading to posterior limits numerically identical to confidence limits.

In fact, with a scalar parameter it is possible in some generality to find a prior giving very close agreement with corresponding confidence intervals. With multidimensional parameters this is not in general possible and naive use of flat priors can lead to procedures that are very poor from all perspectives; see Example 5.5.

An immediate consequence of the Bayesian formulation is that for a given prior distribution and model the data enter only via the likelihood. More specifically for a given prior and model if y and y' are two sets of values with proportional likelihoods the induced posterior distributions are the same. This forms what may be called the *weak likelihood principle*. Less immediately, if two different models but with parameters having the same interpretation, but possibly even referring to different observational systems, lead to data y and y' with proportional likelihoods, then again the posterior distributions are identical. This forms the less compelling *strong likelihood principle*. Most frequentist methods do not obey this latter principle, although the departure is usually relatively minor. If the models refer to different random systems, the implicit prior knowledge may, in any case, be different.

4.3 Frequentist analysis

4.3.1 Extended Fisherian reduction

One approach to simple problems is essentially that of Section 2.5 and can be summarized, as before, in the Fisherian reduction:

- find the likelihood function;
- reduce to a sufficient statistic S of the same dimension as θ;
- find a function of S that has a distribution depending only on ψ;
- place it in pivotal form or alternatively use it to derive p-values for null hypotheses;
- invert to obtain limits for ψ at an arbitrary set of probability levels.

There is sometimes an extension of the method that works when the model is of the (k, d) curved exponential family form. Then the sufficient statistic is of dimension k greater than d, the dimension of the parameter space. We then proceed as follows:

- if possible, rewrite the k-dimensional sufficient statistic, when $k > d$, in the form (S, A) such that S is of dimension d and A has a distribution not depending on θ;
- consider the distribution of S given $A = a$ and proceed as before. The statistic A is called *ancillary*.

There are limitations to these methods. In particular a suitable A may not exist, and then one is driven to asymptotic, i.e., approximate, arguments for problems of reasonable complexity and sometimes even for simple problems.

We give some examples, the first of which is not of exponential family form.

Example 4.1. *Uniform distribution of known range.* Suppose that $(Y_1, \ldots, Y_n)$ are independently and identically distributed in the uniform distribution over $(\theta - 1, \theta + 1)$. The likelihood takes the constant value 2^{-n} provided the smallest and largest values $(y_{(1)}, y_{(n)})$ lie within the range $(\theta - 1, \theta + 1)$ and is zero otherwise. The minimal sufficient statistic is of dimension 2, even though the parameter is only of dimension 1. The model is a special case of a location family and it follows from the invariance properties of such models that $A = Y_{(n)} - Y_{(1)}$ has a distribution independent of θ.

This example shows the imperative of explicit or implicit conditioning on the observed value a of A in quite compelling form. If a is approximately 2, only values of θ very close to $y^* = (y_{(1)} + y_{(n)})/2$ are consistent with the data. If, on the other hand, a is very small, all values in the range of the common observed value y^* plus and minus 1 are consistent with the data. In general, the conditional distribution of Y^* given $A = a$ is found as follows.

The joint density of $(Y_{(1)}, Y_{(n)})$ is

$$n(n-1)(y_{(n)} - y_{(1)})^{n-2}/2^n \tag{4.3}$$

and the transformation to new variables $(Y^*, A = Y_{(n)} - Y_{(1)})$ has unit Jacobian. Therefore the new variables (Y^*, A) have density $n(n-1)a^{(n-2)}/2^n$ defined over the triangular region $(0 \le a \le 2; \theta - 1 + a/2 \le y^* \le \theta + 1 - a/2)$ and density zero elsewhere. This implies that the conditional density of Y^* given $A = a$ is uniform over the allowable interval $\theta - 1 + a/2 \le y^* \le \theta + 1 - a/2$.

Conditional confidence interval statements can now be constructed although they add little to the statement just made, in effect that every value of θ in the relevant interval is in some sense equally consistent with the data. The key point is that an interval statement assessed by its *unconditional* distribution could be formed that would give the correct marginal frequency of coverage but that would hide the fact that for some samples very precise statements are possible whereas for others only low precision is achievable.

Example 4.2. *Two measuring instruments.* A closely related point is made by the following idealized example. Suppose that a single observation Y is made on a normally distributed random variable of unknown mean μ. There are available two measuring instruments, one with known small variance, say $\sigma_0^2 = 10^{-4}$, and one with known large variance, say $\sigma_1^2 = 10^4$. A randomizing device chooses an instrument with probability $1/2$ for each possibility and the full data consist of the observation y and an identifier $d = 0, 1$ to show which variance is involved. The log likelihood is

$$-\log \sigma_d - \exp\{-(y - \mu)^2/(2\sigma_d^2)\}, \tag{4.4}$$

so that (y, d) forms the sufficient statistic and d is ancillary, suggesting again that the formation of confidence intervals or the evaluation of a p-value should use the variance belonging to the apparatus actually used. If the sensitive apparatus,

$d = 0$, is in fact used, why should the interpretation be affected by the possibility that one might have used some other instrument?

There is a distinction between this and the previous example in that in the former the conditioning arose out of the mathematical structure of the problem, whereas in the present example the ancillary statistic arises from a physical distinction between two measuring devices.

There is a further important point suggested by this example. The fact that the randomizing probability is assumed known does not seem material. The argument for conditioning is equally compelling if the choice between the two sets of apparatus is made at random with some unknown probability, provided only that the value of the probability, and of course the outcome of the randomization, is unconnected with μ.

More formally, suppose that we have factorization in which the distribution of S given $A = a$ depends only on θ, whereas the distribution of A depends on an additional parameter γ such that the parameter space becomes $\Omega_\theta \times \Omega_\gamma$, so that θ and γ are variation-independent in the sense introduced in Section 1.1. Then A is ancillary in the extended sense for inference about θ. The term S-ancillarity is often used for this idea.

Example 4.3. *Linear model.* The previous examples are in a sense a preliminary to the following widely occurring situation. Consider the linear model of Examples 1.4 and 2.2 in which the $n \times 1$ vector Y has expectation $E(Y) = z\beta$, where z is an $n \times q$ matrix of full rank $q < n$ and where the components are independently normally distributed with variance σ^2. Suppose that, instead of the assumption of the previous discussion that z is a matrix of fixed constants, we suppose that z is the realized value of a random matrix Z with a known probability distribution.

The log likelihood is by (2.17)

$$-n \log \sigma - \{(y - z\hat{\beta})^T(y - z\hat{\beta}) + (\hat{\beta} - \beta)^T(z^T z)(\hat{\beta} - \beta)\}/(2\sigma^2) \quad (4.5)$$

plus in general functions of z arising from the known distribution of Z. Thus the minimal sufficient statistic includes the residual sum of squares and the least squares estimates as before but also functions of z, in particular $z^T z$. Thus conditioning on z, or especially on $z^T z$, is indicated. This matrix, which specifies the precision of the least squares estimates, plays the role of the distinction between the two measuring instruments in the preceding example.

As noted in the previous discussion and using the extended definition of ancillarity, the argument for conditioning is unaltered if Z has a probability distribution $f_Z(z; \gamma)$, where γ and the parameter of interest are variation-independent.

In many experimental studies the explanatory variables would be chosen by the investigator systematically and treating z as a fixed matrix is totally appropriate. In observational studies in which study individuals are chosen in a random way all variables are random and so modelling z as a random variable might seem natural. The discussion of extended ancillarity shows when that is unnecessary and the standard practice of treating the explanatory variables as fixed is then justified. This does not mean that the distribution of the explanatory variables is totally uninformative about what is taken as the primary focus of concern, namely the dependence of Y on z. In addition to specifying via the matrix $(z^T z)^{-1}$ the precision of the least squares estimates, comparison of the distribution of the components of z with their distribution in some target population may provide evidence of the reliability of the sampling procedure used and of the security of any extrapolation of the conclusions. In a comparable Bayesian analysis the corresponding assumption would be the stronger one that the prior densities of θ and γ are independent.

4.4 Some more general frequentist developments

4.4.1 Exponential family problems

Some limited but important classes of problems have a formally exact solution within the frequentist approach. We begin with two situations involving exponential family distributions. For simplicity we suppose that the parameter ψ of interest is one-dimensional.

We start with a full exponential family model in which ψ is a linear combination of components of the canonical parameter. After linear transformation of the parameter and canonical statistic we can write the density of the observations in the form

$$m(y) \exp\{s_\psi \psi + s_\lambda^T \lambda - k(\psi, \lambda)\}, \tag{4.6}$$

where (s_ψ, s_λ) is a partitioning of the sufficient statistic corresponding to the partitioning of the parameter. For inference about ψ we prefer a pivotal distribution not depending on λ. It can be shown via the mathematical property of completeness, essentially that the parameter space is rich enough, that separation from λ can be achieved only by working with the conditional distribution of the data given $S_\lambda = s_\lambda$. That is, we evaluate the conditional distribution of S_ψ given $S_\lambda = s_\lambda$ and use the resulting distribution to find which values of ψ are consistent with the data at various levels.

Many of the standard procedures dealing with simple problems about Poisson, binomial, gamma and normal distributions can be found by this route.

4.4 Some more general frequentist developments

We illustrate the arguments sketched above by a number of problems connected with binomial distributions, noting that the canonical parameter of a binomial distribution with probability π is $\phi = \log\{\pi/(1-\pi)\}$. This has some advantages in interpretation and also disadvantages. When π is small, the parameter is essentially equivalent to $\log \pi$ meaning that differences in canonical parameter are equivalent to ratios of probabilities.

Example 4.4. *Two-by-two contingency table.* Here there are data on n individuals for each of which two binary features are available, for example blue or nonblue eyes, dead or survived, and so on. Throughout random variables referring to different individuals are assumed to be mutually independent. While this is in some respects a very simple form of data, there are in fact at least four distinct models for this situation. Three are discussed in the present section and one, based on randomization used in design, is deferred to Chapter 9.

These models lead, to some extent, to the same statistical procedures, at least for examining the presence of a relation between the binary features, although derived by slightly different arguments. From an interpretational point of view, however, the models are different. The data are in the form of four frequencies, summing to n, and arranged in a 2×2 array.

In the first place all the R_{kl} in Table 4.1 are random variables with observed values r_{kl}. In some of the later developments one or more marginal frequencies are regarded as fixed and we shall adapt the notation to reflect that.

It helps to motivate the following discussion to consider some ways in which such data might arise. First, suppose that R_{kl} represents the number of occurrences of randomly occurring events in a population over a specified period and that $k = 0, 1$ specifies the gender of the individual and $l = 0, 1$ specifies presence or absence of a risk factor for the event. Then a reasonable model is that the R_{kl} are independent random variables with Poisson distributions. It follows that also the total frequency $R_{..}$ has a Poisson distribution.

Next, suppose that the rows and columns correspond to two separate features of individuals to be treated on an equal footing, say eye colour and hair colour,

Table 4.1. *General notation for* 2×2 *contingency table*

| R_{00} | R_{01} | $R_{0.}$ |
R_{10}	R_{11}	$R_{1.}$
$R_{.0}$	$R_{.1}$	$R_{..} = n$

both regarded as binary for simplicity. Suppose that n independent individuals are chosen and the two features recorded for each. Then the frequencies in Table 4.1 have a multinomial distribution with four cells.

Finally, suppose that the rows of the table correspond to two treatments and that $n_{k.}$ individuals are allocated to treatment k for $k = 0, 1$ and a binary outcome variable, failure or success, say, recorded. Then R_{kl} is the number of responses l among the individuals allocated to treatment k.

In the first situation the objective is to compare the Poisson means, in the second to examine possible association between the features and in the third to compare two probabilities of success. The essential equivalence of the analytical procedures in these situations arises out of successive processes of conditioning; the interpretations, concerning as they do different specifications, are different.

In the first model the R_{kl} are independent random variables with Poisson distributions of means μ_{kl} and canonical parameters $\phi_{kl} = \log \mu_{kl}$. A possible hypothesis of interest is that the Poisson means have multiplicative form, i.e., the canonical parameter is a row effect plus a column effect. To test departures from this model involves the parameter

$$\psi = \log \mu_{11} - \log \mu_{10} - \log \mu_{01} + \log \mu_{00} \qquad (4.7)$$

which is zero in the multiplicative case. The log likelihood may be written in exponential family form and the canonical parameters, $\log \mu$, may be replaced by ψ and three nuisance parameters. For inference about ψ via a distribution not depending on the nuisance parameters, it is necessary to condition on the marginal totals $(R_{0.}, R_{1.}, R_{.0}, R_{.1})$, implying also conditioning on the total number of observations $R_{..}$. The full conditional distribution will be given later in this discussion.

Note first, however, that the marginal distribution of $R_{..}$ is a Poisson distribution of mean $\Sigma \mu_{kl}$. Therefore the conditional distribution of the R_{kl} given $R_{..} = n$ is

$$\frac{\Pi \exp(-\mu_{kl}) \mu_{kl}^{r_{kl}} / r_{kl}!}{\exp(-\Sigma \mu_{kl})(\Sigma \mu_{kl})^n / n!}, \qquad (4.8)$$

so that the conditional distribution has the multinomial form

$$\frac{n!}{\Pi r_{kl}!} \Pi \pi_{kl}^{r_{kl}}, \qquad (4.9)$$

for $r_{..} = \Sigma r_{kl} = n$, say, and is zero otherwise. Here

$$\pi_{kl} = \mu_{kl} / \Sigma \mu_{st}. \qquad (4.10)$$

Further the standard cross-product ratio measure of association $(\pi_{11}\pi_{00})/(\pi_{10}\pi_{01})$ is equal to e^{ψ}. The parameter space is now three-dimensional in the light of the constraint $\Sigma \pi_{kl} = 1$.

4.4 Some more general frequentist developments

Now suppose as a second possibility that a sample of preassigned size n is taken and interest focuses on the joint distribution of two binary features treated on an equal footing. The multinomial distribution just calculated would now be the initial model for analysis and ψ would be a suitable measure of association. Examination of the likelihood again shows that a distribution depending only on ψ is obtained by conditioning on the row and column totals.

Further, if we condition first on the row totals the resulting distribution corresponds to two independent binomial distributions with effective sample sizes $n_{0.} = r_{0.}$ and $n_{1.} = r_{1.}$. The corresponding binomial probabilities are $\pi_{k.} = \pi_{k1}/(\pi_{k0} + \pi_{k1})$ for $k = 0, 1$. The corresponding canonical parameters are

$$\log\{\pi_{k.}/(1 - \pi_{k.})\} = \log(\pi_{k1}/\pi_{k0}). \tag{4.11}$$

Thus the difference of canonical parameters is ψ.

In the preceding formulation there is a sample of size n in which two binary features are recorded and treated on an equal footing. Now suppose that instead the rows of the table represent an explanatory feature, such as alternative treatments and that samples of size $n_{0.} = r_{0.}$ and $n_{1.} = r_{1.}$, say, are taken from the relevant binomial distributions, which have corresponding probabilities of outcome 1 equal to $\pi_{0.}$ and $\pi_{1.}$. The likelihood is the product of two binomial terms

$$\prod \frac{n_{k.}!}{r_{k0}! r_{k1}!} \pi_{k.}^{r_{k0}} (1 - \pi_{k.})^{r_{k1}}. \tag{4.12}$$

On writing this in exponential family form and noting, as above, that ψ is a difference of canonical parameters, we have that a distribution for inference about ψ is obtained by conditioning further on the column totals $R_{.0}$ and $R_{.1}$, so that the distribution is that of, say, R_{11} given the column totals, the row totals already being fixed by design. That is, conditioning is now on the full set $\mathcal{M}$ of margins, namely $R_{.0}, R_{.1}, R_{0.}, R_{1.}$.

This conditional distribution can be written

$$P(R_{11} = r_{11} \mid \mathcal{M}) = \frac{C_{r_{01}}^{r_{0.}} C_{r_{11}}^{r_{1.}} e^{r_{11}\psi}}{\Sigma C_{r_{.1}-w}^{r_{0.}} C_{w}^{r_{1.}} e^{w\psi}}. \tag{4.13}$$

Here C_s^t denotes the number of combinations of t items taken s at a time, namely $t!/\{s!(t-s)!\}$. The distribution is a generalized hypergeometric distribution, reducing to an ordinary hypergeometric distribution in the special case $\psi = 0$, when the sum in the denominator becomes $C_{r_{.1}}^{r_{0.}+r_{1.}}$. The significance test based on this is called Fisher's exact test. The mean and variance of the hypergeometric distribution are respectively

$$\frac{r_{1.}r_{.1}}{n}, \quad \frac{r_{0.}r_{1.}r_{.0}r_{.1}}{n^2(n-1)}. \tag{4.14}$$

We thus have three different models respectively concerning Poisson distributions, concerning association in a multinomial distribution and concerning the comparison of binomial parameters that lead after conditioning to the same statistical procedure. It is helpful to distinguish *conditioning by model formulation* from *technical conditioning*. Thus in the comparison of two binomial parameters the conditioning on row totals (sample sizes) is by model formulation, whereas the conditioning on column totals is technical to derive a distribution not depending on a nuisance parameter.

A practical consequence of the above results is that computerized procedures for any of the models can be used interchangeably in this and related more general situations. Due care in interpretation is of course crucial. In particular methods for log Poisson regression, i.e., the analysis of multiplicative models for Poisson means, can be used much more widely than might seem at first sight.

Later, in Example 9.1, we shall have a version in which all the conditioning is introduced by model formulation.

Example 4.5. *Mantel–Haenszel procedure.* Suppose that there are m independent sets of binary data comparing the same two treatments and thus each producing a two-by-two table of the kind analyzed in Example 4.4 in the final formulation. If we constrain the probabilities in the tables only by requiring that all have the same logistic difference ψ, then the canonical parameters in the kth table can be written

$$\phi_{k0} = \alpha_k, \quad \phi_{k1} = \alpha_k + \psi. \tag{4.15}$$

That is, the relevant probabilities of a response 1 in the two rows of the kth table are, respectively,

$$e^{\alpha_k}/(1 + e^{\alpha_k}), \quad e^{\alpha_k + \psi}/(1 + e^{\alpha_k + \psi}). \tag{4.16}$$

It follows from the discussion for a single table that for inference about ψ treating the α_k as arbitrary nuisance parameters, technical conditioning is needed on the column totals of each table separately as well as on the separate row totals which were fixed by design. Note that if further information were available to constrain the α_k in some way, in particular to destroy variation independence, the conditioning requirement would, in principle at least, be changed. The component of the canonical statistic attached to ψ is $\Sigma R_{k,11}$.

It follows that the distribution for inference about ψ is the convolution of m generalized hypergeometric distributions (4.13). To test the null hypothesis $\psi = 0$ the test statistic $\Sigma R_{k,11}$ can be used and has a distribution formed by convoluting the separate hypergeometric distributions. In fact unless both m and the individual tables are in some sense small, a normal approximation based on adding separate means and variances (4.14) will be satisfactory.

The conditional analysis would be identical if starting from a model in which all cell frequencies had independent Poisson distributions with means specified by a suitable product model.

Example 4.6. *Simple regression for binary data.* Suppose that $Y_1, \ldots, Y_n$ are independent binary random variables such that

$$P(Y_k = 1) = e^{\alpha + \beta z_k}/(1 + e^{\alpha + \beta z_k}),$$
$$P(Y_k = 0) = 1/(1 + e^{\alpha + \beta z_k}),$$
(4.17)

where the notation of Example 1.2 has been used to emphasize the connection with simple normal-theory linear regression. The log likelihood can be written in the form

$$\alpha \Sigma y_k + \beta \Sigma y_k z_k - \Sigma \log(1 + e^{\alpha + \beta z_k}),$$
(4.18)

so that the sufficient statistic is $r = \Sigma y_k$, the total number of 1s, and $s = \Sigma y_k z_k$, the sum of the z_k over the individuals with outcome 1. The latter is conditionally equivalent to the least squares regression coefficient of Y on z.

For inference about β we consider the conditional density of S given $\Sigma Y_k = r$, namely

$$c_r(s) e^{\beta s} / \Sigma c_r(v) e^{\beta v},$$
(4.19)

where the combinatorial coefficient $c_r(s)$ can be interpreted as the number (or proportion) of distinct possible samples of size r drawn without replacement from the finite population $\{z_1, \ldots, z_n\}$ and having total s.

In the particular case of testing the null hypothesis that $\beta = 0$ an exact test of significance is obtained by enumeration of the possible samples or via some approximation. This is a generalization of Fisher's exact test for a 2×2 contingency table, the special case where the z_k are themselves binary.

The above examples have concerned inference about a linear combination of the canonical parameters, which can thus be taken as a single component of that parameter. A rather less elegant procedure deals with the ratio of canonical parameters. For this suppose without essential loss of generality that the parameter of interest is $\psi = \phi_1/\phi_2$, the ratio of the first two canonical parameters. Under a null hypothesis $\psi = \psi_0$ the log likelihood is thus

$$\exp\{(s_1 \psi_0 + s_2)\phi_2 + s^{[3]T} \phi^{[3]} - k(\phi)\},$$
(4.20)

where, for example, $\phi^{[3]} = (\phi_3, \phi_4, \ldots)$. It follows that in order to achieve a distribution independent of the nuisance parameters technical conditioning is required on $(s_1\psi_0 + s_2, s^{[3]})$. That is, to test for departures from the null hypothesis leading to, say, large values of S_1, we consider the upper tail of its conditional distribution.

Example 4.7. *Normal mean, variance unknown.* The canonical parameters for a normal distribution of unknown mean μ and unknown variance σ^2 are μ/σ^2 and $1/\sigma^2$. Thus inference about μ is concerned with the ratio of canonical parameters. Under the null hypothesis $\mu = \mu_0$, the sufficient statistic for σ^2 is $\Sigma(Y_k - \mu_0)^2$. Under the null hypothesis the conditional distribution of the vector Y is uniform on the sphere defined by the conditioning statistic and for the observed value $\bar{y}$ the required p-value is the proportion of the surface area of the sphere corresponding to a cap defined by $\bar{y}$ and centred on the unit line through ψ_0. Some calculation, which will not be reproduced here, shows this argument to be a rather tortuous route to the standard Student t statistic and its distribution.

A similar but slightly more complicated argument yields a significance test for a regression parameter in a normal-theory linear model.

Example 4.8. *Comparison of gamma distributions.* Suppose that Y_1, Y_2 are independently distributed with densities

$$\rho_k(\rho_k y_k)^{m_k-1} \exp(-\rho_k y_k)/\Gamma(m_k) \tag{4.21}$$

for $k = 1, 2$ and that interest lies in $\psi = \rho_2/\rho_1$. This might arise in comparing the rates of two Poisson processes or alternatively in considering the ratio of two normal-theory estimates of variance having respectively $(2m_1, 2m_2)$ degrees of freedom.

The general argument leads to consideration of the null hypothesis $\rho_2 = \psi_0 \rho_1$ and thence to study of the conditional distribution of Y_1 given $Y_1 + \psi_0 Y_2$. This can be shown to be equivalent to using the pivot

$$(\rho_2 Y_2/m_2)/(\rho_1 Y_1/m_1) \tag{4.22}$$

having pivotal distribution the standard variance-ratio or F distribution with $(2m_1, 2m_2)$ degrees of freedom.

The reason that in the last two examples a rather round-about route is involved in reaching an apparently simple conclusion is that the exponential family takes no account of simplifications emerging from the transformational structure of scale and/or location intrinsic to these two problems.

Example 4.9. *Unacceptable conditioning.* Suppose that (Y_1, Y_2) have independent Poisson distribution of means (μ_1, μ_2) and hence canonical parameters

4.4 Some more general frequentist developments

$(\log \mu_1, \log \mu_2)$. A null hypothesis about the ratio of canonical parameters, namely $\log \mu_2 = \psi_0 \log \mu_1$, is

$$\mu_2 = \mu_1^{\psi_0}. \tag{4.23}$$

The previous argument requires conditioning the distribution of Y_2 on the value of $y_2 + \psi_0 y_1$. Now especially if ψ_0 is irrational, or even if it is rational of the form a/b, where a and b are large coprime integers, there may be very few points of support in the conditional distribution in question. In an extreme case the observed point may be unique. The argument via technical conditioning may thus collapse; moreover small changes in ψ_0 induce dramatic changes in the conditional distribution. Yet the statistical problem, even if somewhat artificial, is not meaningless and is clearly capable of some approximate answer. If the observed values are not too small the value of ψ cannot be too far from $\log y_2 / \log y_1$ and the arguments of asymptotic theory to be developed later in the book can be deployed to produce approximate tests and confidence intervals.

The important implication of this example is that the use of technical conditioning to produce procedures whose properties do not depend on nuisance parameters is powerful and appealing, but that insistence on *exact* separation from the nuisance parameter may be achieved at too high a price.

4.4.2 Transformation models

A different kind of reduction is possible for models such as the location model which preserve a special structure under a family of transformations.

Example 4.10. *Location model.* We return to Example 1.8, the location model. The likelihood is $\Pi g(y_k - \mu)$ which can be rearranged in terms of the order statistics $y_{(l)}$, i.e., the observed values arranged in increasing order of magnitude. These in general form the sufficient statistic for the model, minimal except for the normal distribution. Now the random variables corresponding to the differences between order statistics, specified, for example, as

$$a = (y_{(2)} - y_{(1)}, \ldots, y_{(n)} - y_{(1)}) = (a_2, \ldots, a_n), \tag{4.24}$$

have a distribution not depending on μ and hence are ancillary; inference about μ is thus based on the conditional distribution of, say, $y_{(1)}$ given $A = a$. The choice of $y_{(1)}$ is arbitrary because any estimate of location, for example the mean, is $y_{(1)}$ plus a function of a, and the latter is not random after conditioning. Now the marginal density of A is

$$\int g(y_{(1)} - \mu) g(y_{(1)} + a_2 - \mu) \cdots g(y_{(1)} + a_n - \mu) dy_{(1)}. \tag{4.25}$$

The integral with respect to $y_{(1)}$ is equivalent to one integrated with respect to μ so that, on reexpressing the integrand in terms of the likelihood for the unordered variables, the density of A is the function necessary to normalize the likelihood to integrate to one with respect to μ. That is, the conditional density of $Y_{(1)}$, or of any other measure of location, is determined by normalizing the likelihood function and regarding it as a function of $y_{(1)}$ for given a. This implies that p-values and confidence limits for μ result from normalizing the likelihood and treating it like a distribution of μ.

This is an exceptional situation in frequentist theory. In general, likelihood is a point function of its argument and integrating it over sets of parameter values is not statistically meaningful. In a Bayesian formulation it would correspond to combination with a uniform prior but no notion of a prior distribution is involved in the present argument.

The ancillary statistics allow testing conformity with the model. For example, to check on the functional form of $g(.)$ it would be best to work with the differences of the ordered values from the mean, $(y_{(l)} - \bar{y})$ for $l = 1, \ldots, n$. These are functions of a as previously defined. A test statistic could be set up informally sensitive to departures in distributional form from those implied by $g(.)$. Or $(y_{(l)} - \bar{y})$ could be plotted against $G^{-1}\{(l - 1/2)/(n+1)\}$, where $G(.)$ is the cumulative distribution function corresponding to the density $g(.)$.

From a more formal point of view, there is defined a set, in fact a group, of transformations on the sample space, namely the set of location shifts. Associated with that is a group on the parameter space. The sample space is thereby partitioned into orbits defined by constant values of a, invariant under the group operation, and position on the orbit conditional on the orbit in question provides the information about μ. This suggests more general formulations of which the next most complicated is the scale and location model in which the density is $\tau^{-1}g\{(y-\mu)/\tau)\}$. Here the family of implied transformations is two-dimensional, corresponding to changes of scale and location and the ancillary statistics are standardized differences of order statistics, i.e.,

$$\{(y_{(3)} - y_{(1)})/(y_{(2)} - y_{(1)}), \ldots, (y_{(n)} - y_{(1)})/(y_{(2)} - y_{(1)})\} = (a_3^*, \ldots, a_n^*). \tag{4.26}$$

There are many essentially equivalent ways of specifying the ancillary statistic.

We can thus write estimates of μ and τ as linear combinations of $y_{(1)}$ and $y_{(2)}$ with coefficients depending on the fixed values of the ancillary statistics. Further, the Jacobian of the transformation from the $y_{(k)}$ to the variables

$(y_{(1)}, y_{(2)}, a^*)$ is $(y_{(2)} - y_{(1)})^{n-2}$, so that all required distributions are directly derivable from the likelihood function.

A different and in some ways more widely useful role of transformation models is to suggest reduction of a sufficient statistic to focus on a particular parameter of interest.

Example 4.11. *Normal mean, variance unknown (ctd).* To test the null hypothesis $\mu = \mu_0$ when the variance σ^2 is unknown using the statistics $(\bar{Y}, s^2)$ we need a function that is unaltered by the transformations of origin and units of measurement encapsulated in

$$(\bar{Y}, s^2) \text{ to } (a + b\bar{Y}, b^2 s^2) \qquad (4.27)$$

and

$$\mu \text{ to } a + b\mu. \qquad (4.28)$$

This is necessarily a function of the standard Student t statistic.

4.5 Some further Bayesian examples

In principle the prior density in a Bayesian analysis is an insertion of additional information and the form of the prior should be dictated by the nature of that evidence. It is useful, at least for theoretical discussion, to look at priors which lead to mathematically tractable answers. One such form, useful usually only for nuisance parameters, is to take a distribution with finite support, in particular a two-point prior. This has in one dimension three adjustable parameters, the position of two points and a probability, and for some limited purposes this may be adequate. Because the posterior distribution remains concentrated on the same two points computational aspects are much simplified.

We shall not develop that idea further and turn instead to other examples of parametric conjugate priors which exploit the consequences of exponential family structure as exemplified in Section 2.4.

Example 4.12. *Normal variance.* Suppose that $Y_1, \ldots, Y_n$ are independently normally distributed with known mean, taken without loss of generality to be zero, and unknown variance σ^2. The likelihood is, except for a constant factor,

$$\frac{1}{\sigma^n} \exp\{-\Sigma y_k^2/(2\sigma^2)\}. \qquad (4.29)$$

The canonical parameter is $\phi = 1/\sigma^2$ and simplicity of structure suggests taking ϕ to have a prior gamma density which it is convenient to write in the form

$$\pi(\phi; g, n_\pi) = g(g\phi)^{n_\pi/2-1} e^{-g\phi}/\Gamma(n_\pi/2), \qquad (4.30)$$

defined by two quantities assumed known. One is n_π, which plays the role of an effective sample size attached to the prior density, by analogy with the form of the chi-squared density with n degrees of freedom. The second defining quantity is g. Transformed into a distribution for σ^2 it is often called the *inverse gamma distribution*. Also $E_\pi(\Phi) = n_\pi/(2g)$.

On multiplying the likelihood by the prior density, the posterior density of Φ is proportional to

$$\phi^{(n+n_\pi)/2-1} \exp\left[-\{(\Sigma y_k^2 + n_\pi/E_\pi(\Phi))\}\phi/2\right]. \tag{4.31}$$

The posterior distribution is in effect found by treating

$$\{\Sigma y_k^2 + n_\pi/E_\pi(\Phi)\}\Phi \tag{4.32}$$

as having a chi-squared distribution with $n + n_\pi$ degrees of freedom.

Formally, frequentist inference is based on the pivot $\Sigma Y_k^2/\sigma^2$, the pivotal distribution being the chi-squared distribution with n degrees of freedom. There is formal, although of course not conceptual, equivalence between the two methods when $n_\pi = 0$. This arises from the improper prior $d\phi/\phi$, equivalent to $d\sigma/\sigma$ or to a uniform improper prior for $\log \sigma$. That is, while there is never in this setting exact agreement between Bayesian and frequentist solutions, the latter can be approached as a limit as $n_\pi \to 0$.

Example 4.13. *Normal mean, variance unknown (ctd).* In Section 1.5 and Example 4.12 we have given simple Bayesian posterior distributions for the mean of a normal distribution with variance known and for the variance when the mean is known.

Suppose now that both parameters are unknown and that the mean is the parameter of interest. The likelihood is proportional to

$$\frac{1}{\sigma^n} \exp[-\{n(\bar{y} - \mu)^2 + \Sigma(y_k - \bar{y})^2\}/(2\sigma^2)]. \tag{4.33}$$

In many ways the most natural approach may seem to be to take the prior distributions for mean and variance to correspond to independence with the forms used in the single-parameter analyses. That is, an inverse gamma distribution for σ^2 is combined with a normal distribution of mean m and variance v for μ. We do not give the details.

A second possibility is to modify the above assessment by replacing v by $b\sigma^2$, where b is a known constant. Mathematical simplification that results from this stems from μ/σ^2 rather than μ itself being a component of the canonical parameter. A statistical justification might sometimes be that σ establishes the natural scale for measuring random variability and that the amount of prior

4.5 Some further Bayesian examples

information is best assessed relative to that. Prior independence of mean and variance has been abandoned.

The product of the likelihood and the prior is then proportional to

$$\phi^{n/2+n_\pi/2-1} e^{-w\phi}, \qquad (4.34)$$

where

$$w = \mu^2\{n/2 + 1/(2b)\} - \mu(n\bar{y} + m/b) + g \\ + n\bar{y}^2/2 + \Sigma(y_k - \bar{y})^2/2 + m^2/(2b). \qquad (4.35)$$

To obtain the marginal posterior distribution of μ we integrate with respect to ϕ and then normalize the resulting function of μ. The result is proportional to $w^{-n/2-n_\pi/2}$. Now note that the standard Student t distribution with d degrees of freedom has density proportional to $(1 + t^2/d)^{-d(d+1)/2}$. Comparison with w shows without detailed calculation that the posterior density of μ involves the Student t distribution with degrees of freedom $n + n_\pi - 1$. More detailed calculation shows that $(\mu - \tilde{\mu}_\pi)/\tilde{s}_\pi$ has a posterior t distribution, where

$$\tilde{\mu}_\pi = (n\bar{y} + m/b)/(n + 1/b) \qquad (4.36)$$

is the usual weighted mean and $\tilde{s}_\pi$ is a composite estimate of precision obtained from $\Sigma(y_k - \bar{y})^2$, the prior estimate of variance and the contrast $(m - \bar{y})^2$.

Again the frequentist solution based on the pivot

$$\frac{(\mu - \bar{Y})\sqrt{n}}{\{\Sigma(Y_k - \bar{Y})^2/(n-1)\}^{1/2}}, \qquad (4.37)$$

the standard Student t statistic with degrees of freedom $n - 1$, is recovered but only as a limiting case.

Example 4.14. *Components of variance.* We now consider an illustration of a wide class of problems introduced as Example 1.9, in which the structure of unexplained variation in the data is built up from more than one component random variable. We suppose that the observed random variable Y_{ks} for $k = 1, \ldots, m; s = 1, \ldots, t$ has the form

$$Y_{ks} = \mu + \eta_k + \epsilon_{ks}, \qquad (4.38)$$

where μ is an unknown mean and the η and the ϵ are mutually independent normally distributed random variables with zero mean and variances, respectively τ_η, τ_ϵ, called components of variance.

Suppose that interest focuses on the components of variance. To simplify some details suppose that μ is known. In the balanced case studied here, and only then, there is a reduction to a (2, 2) exponential family with canonical parameters $(1/\tau_\epsilon, 1/(\tau_\epsilon + t\tau_\eta))$ so that a simple analytical form would emerge from giving these two parameters independent gamma prior distributions,

While this gives an elegant result, the dependence on the subgroup size t is for most purposes likely to be unreasonable. The restriction to balanced problems, i.e., those in which the subgroup size t is the same for all k is also limiting, at least so far as exact analytic results are concerned.

A second possibility is to assign the separate variance components independent inverse gamma priors and a third is to give one variance component, say τ_ϵ, an inverse gamma prior and to give the ratio τ_η/τ_ϵ an independent prior distribution. There is no difficulty in principle in evaluating the posterior distribution of the unknown parameters and any function of them of interest.

An important more practical point is that the relative precision for estimating the upper variance component τ_η is less, and often much less, than that for estimating a variance from a group of m observations and having $m-1$ degrees of freedom in the standard terminology. Thus the inference will be more sensitive to the form of the prior density than is typically the case in simpler examples. The same remark applies even more strongly to complex multilevel models; it is doubtful whether such models should be used for a source of variation for which the effective replication is less than 10–20.

Notes 4

Section 4.2. For a general discussion of nonpersonalistic Bayesian theory and of a range of statistical procedures developed from this perspective, see Box and Tiao (1973). Box (1980) proposes the use of Bayesian arguments for parametric inference and frequentist arguments for model checking. Bernardo and Smith (1994) give an encyclopedic account of Bayesian theory largely from the personalistic perspective. See also Lindley (1990). For a brief historical review, see Appendix A.

Bayesian approaches are based on formal requirements for the concept of probability. Birnbaum (1962) aimed to develop a concept of statistical evidence on the basis of a number of conditions that the concept should satisfy and it was argued for a period that this approach also led to essentially Bayesian methods. There were difficulties, however, probably in part at least because of confusions between the weak and strong likelihood principles and in his final paper Birnbaum (1969) considers sensible confidence intervals to be the primary mode of inference. Fisher (1956), by contrast, maintains that different contexts require different modes of inference, in effect arguing against any single unifying theory.

Section 4.3. The discussion of the uniform distribution follows Welch (1939), who, however, drew and maintained the conclusion that the unconditional distribution is appropriate for inference. The problem of the two measurement instruments was introduced by Cox (1958c) in an attempt to clarify the issues involved. Most treatments of regression and the linear model adopt the conditional formulation of Example 4.3, probably wisely without comment.

Section 4.4. The relation between the Poisson distribution and the multinomial distribution was used as a mathematical device by R. A. Fisher in an early discussion of degrees of freedom in the testing of, for example, contingency tables. It is widely used in the context of log Poisson regression in the analysis of epidemiological data. The discussion of the Mantel–Haenszel and related problems follows Cox (1958a, 1958b) and Cox and Snell (1988). The discussion of transformation models stems from Fisher (1934) and is substantially developed by Fraser (1979) and forms the basis of much of his subsequent work (Fraser, 2004), notably in approximating other models by models locally with transformation structure.

Completeness of a family of distributions means essentially that there is no nontrivial function of the random variable whose expectation is zero for all members of the family. The notion is discussed in the more mathematical of the books listed in Notes 1.

In the Neyman–Pearson development the elimination of nuisance parameters is formalized in the notions of similar tests and regions of Neyman structure (Lehmann and Romano, 2004).

The distinction between the various formulations of 2×2 contingency tables is set out with great clarity by Pearson (1947). Fisher (1935b) introduces the generalized hypergeometric distribution with the somewhat enigmatic words: 'If it be admitted that these marginal frequencies by themselves supply no information on the point at issue ...'. Barnard introduced an unconditional procedure that from a Neyman–Pearson perspective achieved a test of preassigned probability under H_0 and improved properties under alternatives, but withdrew the procedure, after seeing a comment by Fisher. The controversy continues. From the perspective taken here achieving a preassigned significance level is not an objective and the issue is more whether some form of approximate conditioning might be preferable.

5
Interpretations of uncertainty

Summary. This chapter discusses the nature of probability as it is used to represent both variability and uncertainty in the various approaches to statistical inference. After some preliminary remarks, the way in which a frequency notion of probability can be used to assess uncertainty is reviewed. Then two contrasting notions of probability as representing degree of belief in an uncertain event or hypothesis are examined.

5.1 General remarks

We can now consider some issues involved in formulating and comparing the different approaches.

In some respects the Bayesian formulation is the simpler and in other respects the more difficult. Once a likelihood and a prior are specified to a reasonable approximation all problems are, in principle at least, straightforward. The resulting posterior distribution can be manipulated in accordance with the ordinary laws of probability. The difficulties centre on the concepts underlying the definition of the probabilities involved and then on the numerical specification of the prior to sufficient accuracy.

Sometimes, as in certain genetical problems, it is reasonable to think of θ as generated by a stochastic mechanism. There is no dispute that the Bayesian approach is at least part of a reasonable formulation and solution in such situations. In other cases to use the formulation in a literal way we have to regard probability as measuring uncertainty in a sense not necessarily directly linked to frequencies. We return to this issue later. Another possible justification of some Bayesian methods is that they provide an algorithm for extracting from the likelihood some procedures whose fundamental strength comes from frequentist considerations. This can be regarded, in particular, as supporting a broad class of procedures, known as shrinkage methods, including ridge regression.

The emphasis in this book is quite often on the close relation between answers possible from different approaches. This does not imply that the different views never conflict. Also the differences of interpretation between different numerically similar answers may be conceptually important.

5.2 Broad roles of probability

A recurring theme in the discussion so far has concerned the broad distinction between the frequentist and the Bayesian formalization and meaning of probability. Kolmogorov's axiomatic formulation of the theory of probability largely decoupled the issue of meaning from the mathematical aspects; his axioms were, however, firmly rooted in a frequentist view, although towards the end of his life he became concerned with a different interpretation based on complexity. But in the present context meaning is crucial.

There are two ways in which probability may be used in statistical discussions. The first is phenomenological, to describe in mathematical form the empirical regularities that characterize systems containing haphazard variation. This typically underlies the formulation of a probability model for the data, in particular leading to the unknown parameters which are regarded as a focus of interest. The probability of an event $\mathcal{E}$ is an idealized limiting proportion of times in which $\mathcal{E}$ occurs in a large number of repeat observations on the system under the same conditions. In some situations the notion of a large number of repetitions can be reasonably closely realized; in others, as for example with economic time series, the notion is a more abstract construction. In both cases the working assumption is that the parameters describe features of the underlying data-generating process divorced from special essentially accidental features of the data under analysis.

That first phenomenological notion is concerned with describing *variability*. The second role of probability is in connection with *uncertainty* and is thus epistemological. In the frequentist theory we adapt the frequency-based view of probability, using it only indirectly to calibrate the notions of confidence intervals and significance tests. In most applications of the Bayesian view we need an extended notion of probability as measuring the uncertainty of $\mathcal{E}$ given $\mathcal{F}$, where now $\mathcal{E}$, for example, is not necessarily the outcome of a random system, but may be a hypothesis or indeed any feature which is unknown to the investigator. In statistical applications $\mathcal{E}$ is typically some statement about the unknown parameter θ or more specifically about the parameter of interest ψ. The present discussion is largely confined to such situations. The issue of whether a single number could usefully encapsulate uncertainty about the correctness of, say, the Fundamental Theory underlying particle physics is far outside the scope of the present discussion. It could, perhaps, be applied to a more specific question such as a prediction of the Fundamental Theory: will the Higgs boson have been discovered by 2010?

One extended notion of probability aims, in particular, to address the point that in interpretation of data there are often sources of uncertainty additional to those arising from narrow-sense statistical variability. In the frequentist approach these aspects, such as possible systematic errors of measurement, are addressed qualitatively, usually by formal or informal sensitivity analysis, rather than incorporated into a probability assessment.

5.3 Frequentist interpretation of upper limits

First we consider the frequentist interpretation of upper limits obtained, for example, from a suitable pivot. We take the simplest example, Example 1.1, namely the normal mean when the variance is known, but the considerations are fairly general. The upper limit

$$\bar{y} + k_c^* \sigma_0 / \sqrt{n}, \qquad (5.1)$$

derived here from the probability statement

$$P(\mu < \bar{Y} + k_c^* \sigma_0 / \sqrt{n}) = 1 - c, \qquad (5.2)$$

is a particular instance of a *hypothetical* long run of statements a proportion $1 - c$ of which will be true, always, of course, assuming our model is sound. We can, at least in principle, make such a statement for each c and thereby generate a collection of statements, sometimes called a *confidence distribution*. There is no restriction to a single c, so long as some compatibility requirements hold.

Because this has the formal properties of a distribution for μ it was called by R. A. Fisher the *fiducial distribution* and sometimes the *fiducial probability distribution*. A crucial question is whether this distribution can be interpreted and manipulated like an ordinary probability distribution. The position is:

- a single set of limits for μ from some data can in some respects be considered just like a probability statement for μ;
- such probability statements cannot in general be combined or manipulated by the laws of probability to evaluate, for example, the chance that μ exceeds some *given* constant, for example zero. This is clearly illegitimate in the present context.

5.3 Frequentist interpretation of upper limits

That is, as a single statement a $1 - c$ upper limit has the evidential force of a statement of a unique event within a probability system. But the rules for manipulating probabilities in general do not apply. The limits are, of course, directly based on probability calculations.

Nevertheless the treatment of the confidence interval statement about the parameter as if it is *in some respects* like a probability statement contains the important insights that, in inference for the normal mean, the unknown parameter is more likely to be near the centre of the interval than near the end-points and that, provided the model is reasonably appropriate, if the mean is outside the interval it is not likely to be far outside.

A more emphatic demonstration that the sets of upper limits defined in this way do not determine a probability distribution is to show that in general there is an inconsistency if such a formal distribution determined in one stage of analysis is used as input into a second stage of probability calculation. We shall not give details; see Note 5.2.

The following example illustrates in very simple form the care needed in passing from assumptions about the data, given the model, to inference about the model, given the data, and in particular the false conclusions that can follow from treating such statements as probability statements.

Example 5.1. *Exchange paradox.* There are two envelopes, one with ϕ euros and the other with 2ϕ euros. One is chosen at random and given to you and when opened it is found to contain 10^3 euros. You now have a choice. You may keep the 10^3 euros or you may open the other envelope in which case you keep its contents. You argue that the other envelope is equally likely to contain 500 euros or 2×10^3 euros and that, provided utility is linear in money, the expected utility of the new envelope is 1.25×10^3 euros. Therefore you take the new envelope. This decision does not depend on the particular value 10^3 so that there was no need to open the first envelope.

The conclusion is clearly wrong. The error stems from attaching a probability in a non-Bayesian context to the content of the new and as yet unobserved envelope, given the observation; the only probability initially defined is that of the possible observation, given the parameter, i.e., given the content of both envelopes. Both possible values for the content of the new envelope assign the same likelihood to the data and are in a sense equally consistent with the data. In a non-Bayesian setting, this is not the same as the data implying equal *probabilities* to the two possible values. In a Bayesian discussion with a proper prior distribution, the relative prior probabilities attached to 500 euros and 2000 euros would be crucial. The improper prior that attaches equal prior probability to all values of ϕ is trapped by the same paradox as that first formulated.

5.4 Neyman–Pearson operational criteria

We now outline a rather different approach to frequentist intervals not via an initial assumption that the dependence on the data is dictated by consideration of sufficiency. Suppose that we wish to find, for example, upper limits for ψ with the appropriate frequency interpretation, i.e., derived from a property such as

$$P(\mu < \bar{Y} + k_c^* \sigma_0/\sqrt{n}) = 1 - c. \tag{5.3}$$

Initially we may require that exactly (or to an adequate approximation) for all θ

$$P\{\psi < T(Y; c)\} = 1 - c, \tag{5.4}$$

where $T(Y; c)$ is a function of the observed random variable Y. This ensures that the right coverage probability is achieved. There are often many ways of achieving this, some in a sense more efficient than others. It is appealing to define optimality, i.e., sensitivity of the analysis, by requiring

$$P\{\psi' < T(Y; c)\} \tag{5.5}$$

to be minimal for all $\psi' > \psi$ subject to correct coverage. This expresses the notion that the upper bound should be as small as possible and does so in a way that is essentially unchanged by monotonic increasing transformations of ψ and T.

Essentially this strong optimality requirement is satisfied only when a simple Fisherian reduction is possible. In the Neyman–Pearson approach dependence on the sufficient statistic, in the specific example $\bar{Y}$, emerges out of the optimality requirements, rather than being taken as a starting point justified by the defining property of sufficient statistics.

5.5 Some general aspects of the frequentist approach

The approach via direct study of hypothetical long-run frequency properties has the considerable advantage that it provides a way of comparing procedures that may have compensating properties of robustness, simplicity and directness and of considering the behaviour of procedures when the underlying assumptions are not satisfied.

There remains an important issue, however. Is it clear in each case what is the relevant long run of repetitions appropriate for the interpretation of the specific data under analysis?

Example 5.2. *Two measuring instruments (ctd).* The observed random variables are (Y, D), where Y represents the measurement and D is an indicator of which instrument is used, i.e., we know which variance applies to our observation.

The minimal sufficient statistic is (Y, D) with an obvious generalization if repeat observations are made on the same apparatus. We have a (2, 1) exponential family in which, however, one component of the statistic, the indicator D, has only two points of support. By the Fisherian reduction we condition on the observed and exactly known variance, i.e., we use the normal distribution we know was determined by the instrument actually used and we take no account of the fact that in repetitions we might have obtained a quite different precision.

Unfortunately the Neyman–Pearson approach does not yield this result automatically; it conflicts, superficially at least, with the sensitivity requirement for optimality. That is, if we require a set of sample values leading to the long-run inclusion of the true value with a specified probability and having maximum probability of exclusion of (nearly all) false values, the conditional procedure is not obtained.

We need a supplementary principle to define the appropriate ensemble of repetitions that determines the statistical procedure. Note that this is necessary even if the repetitive process were realized physically, for example if the above measuring procedure took place every day over a long period.

This example can be regarded as a pared-down version of Example 4.3, concerning regression analysis. The use of D to define a conditional distribution is at first sight conceptually different from that used to obtain inference free of a nuisance parameter.

5.6 Yet more on the frequentist approach

There are two somewhat different ways of considering frequentist procedures. The one in some sense closest to the measurement of strength of evidence is the calculation of p-values as a measuring instrument. This is to be assessed, as are other measuring devices, by calibration against its performance when used. The calibration is the somewhat hypothetical one that a given p has the interpretation given previously in Section 3.2. That is, were we to regard a particular p as just decisive against the null hypothesis, then in only a long-run proportion p of cases in which the null hypothesis is true would it be wrongly rejected. In this view such an interpretation serves only to explain the meaning of a particular p.

A more direct view is to regard a significance test at some preassigned level, for example 0.05, as defining a procedure which, if used as a basis for interpretation, controls a certain long-run error rate. Then, in principle at least, a critical threshold value of p should be chosen in the context of each application and used as indicated. This would specify what is sometimes called a *rule of inductive behaviour*. If taken very literally no particular statement of a probabilistic nature is possible about any specific case. In its extreme form, we report a departure from the null hypothesis if and only if p is less than or equal to the preassigned level. While procedures are rarely used in quite such a mechanical form, such a procedure is at least a crude model of reporting practice in some areas of science and, while subject to obvious criticisms and improvement, does give some protection against the overemphasizing of results of transitory interest. The procedure is also a model of the operation of some kinds of regulatory agency.

We shall not emphasize that approach here but concentrate on the reporting of p-values as an analytical and interpretational device. In particular, this means that any probabilities that are calculated are to be relevant to what are sometimes referred to as the unique set of data under analysis.

Example 5.3. *Rainy days in Gothenburg.* Consider daily rainfall measured at some defined point in Gothenburg, say in April. To be specific let W be the event that on a given day the rainfall exceeds 5 mm. Ignore climate change and any other secular changes and suppose we have a large amount of historical data recording the occurrence and nonoccurrence of W on April days. For simplicity, ignore possible dependence between nearby days. The relative frequency of W when we aggregate will tend to stabilize and we idealize this to a limiting value π_W, the probability of a wet day. This is frequency-based, a physical measure of the weather-system and what is sometimes called a *chance* to distinguish it from probability as assessing uncertainty.

Now consider the question: will it be wet in Gothenburg tomorrow, a very specific day in April?

Suppose we are interested in this unique day, not in a sequence of predictions. If probability is a degree of belief, then there are arguments that in the absence of other evidence the value π_W should apply or if there are other considerations, then they have to be combined with π_W.

But supposing that we stick to a frequentist approach. There are then two lines of argument. The first, essentially that of a rule of inductive behaviour, is that probability is inapplicable (at least until tomorrow midnight by when the probability will be either 0 or 1). We may follow the rule of saying that 'It will rain tomorrow'. If we follow such a rule repeatedly we will be right a proportion π_W of times but no further statement about tomorrow is possible.

Another approach is to say that the probability for tomorrow is π_W but only if two conditions are satisfied. The first is that tomorrow is a member of an ensemble of repetitions in a proportion π_W of which the event occurs.

5.7 Personalistic probability

The second is that one cannot establish tomorrow as a member of a sub-ensemble with a different proportion, that tomorrow is not a member of a *recognizable subset*. That is, the statement must be adequately conditional. There are substantial difficulties in implementing this notion precisely in a mathematical formalization and these correspond to serious difficulties in statistical analysis. If we condition too far, every event is unique. Nevertheless the notion of appropriate conditioning captures an essential aspect of frequency-based inference.

Example 5.4. *The normal mean (ctd).* We now return to the study of limits for the mean μ of a normal distribution with known variance; this case is taken purely for simplicity and the argument is really very general. In the Bayesian discussion we derive a distribution for μ which is, when the prior is "flat", normal with mean $\bar{y}$ and variance σ_0^2/n. In the frequentist approach we start from the statement

$$P(\mu < \bar{Y} + k_c^* \sigma_0/\sqrt{n}) = 1 - c. \qquad (5.6)$$

Then we take our data with mean $\bar{y}$ and substitute into the previous equation to obtain a limit for μ, namely $\bar{y} + p_c \sigma_0/\sqrt{n}$. Following the discussion of Example 5.3 we have two interpretations. The first is that probability does not apply, only the properties of the rule of inductive behaviour. The second is that probability does apply, provided there is no further conditioning set available that would lead to a (substantially) different probability. Unfortunately, while at some intuitive level it is clear that, in the absence of further information, no useful further conditioning set is available, careful mathematical discussion of the point is delicate and the conclusions less clear than one might have hoped; see Note 5.6.

5.7 Personalistic probability

We now turn to approaches that attempt to measure uncertainty directly by probability and which therefore potentially form a basis for a general Bayesian discussion.

A view of probability that has been strongly advocated as a systematic base for statistical theory and analysis is that $P(\mathcal{E} \mid \mathcal{F})$ represents the degree of belief in $\mathcal{E}$, given $\mathcal{F}$, held by a particular individual, commonly referred to as You. It is sometimes convenient to omit the conditioning information $\mathcal{F}$ although in applications it is always important and relevant and includes assumptions, for example, about the data-generating process which You consider sensible.

While many treatments of this topic regard the frequency interpretation of probability merely as the limiting case of Your degree of belief when there is a large amount of appropriate data, it is in some ways simpler in defining $P(\mathcal{E})$ to suppose available in principle for any p such that $0 < p < 1$ an agreed system of generating events with long-run frequency of occurrence p, or at least generally agreed to have probability p. These might, for example, be based on a well-tested random number generator.

You then consider a set of hypothetical situations in which You must choose between

- a valuable prize if $\mathcal{E}$ is true and zero otherwise, or
- that same prize if one outcome of the random system with probability p occurs and zero otherwise.

The value of p at which You are indifferent between these choices is defined to be $P(\mathcal{E})$; such a p exists under weak conditions. The argument implies the existence of $P(\mathcal{E})$ in principle for *all* possible events $\mathcal{E}$. The definition is not restricted to events in replicated random systems.

In statistical arguments the idea is in principle applied for each value of θ^* of the parameter to evaluate the probability that the parameter Θ, considered now as random, is less than θ^*, i.e., in principle to build the prior distribution of Θ.

To show that probability as so defined has the usual mathematical properties of probability, arguments can be produced to show that behaviour in different but related situations that was inconsistent with the standard laws of probability would be equivalent to making loss-making choices in the kind of situation outlined above. Thus the discussion is not necessarily about the behaviour of real people but rather is about the internal consistency of Your probability assessments, so-called *coherency*. In fact experiments show that coherency is often not achieved, even in quite simple experimental settings.

Because Your assessments of uncertainty obey the laws of probability, they satisfy Bayes' theorem. Hence it is argued that the only coherent route to statistical assessment of Your uncertainty given the data y must be consistent with the application of Bayes' theorem to calculate the posterior distribution of the parameter of interest from the prior distribution and the likelihood given by the model.

In this approach, it is required that participation in the choices outlined above is compulsory; there is no concept of being ignorant about $\mathcal{E}$ and of refusing to take part in the assessment.

Many treatments of this type of probability tie it strongly to the decisions You have to make on the basis of the evidence under consideration.

5.8 Impersonal degree of belief

It is explicit in the treatment of the previous section that the probabilities concerned belong to a particular person, You, and there is no suggestion that even given the same information $\mathcal{F}$ different people will have the same probability. Any agreement between individuals is presumed to come from the availability of a large body of data with an agreed probability model, when the contribution of the prior will often be minor, as indeed we have seen in a number of examples.

A conceptually quite different approach is to define $P(\mathcal{E} \mid \mathcal{F})$ as the degree of belief in $\mathcal{E}$ that a reasonable person would hold given $\mathcal{F}$. The presumption then is that differences between reasonable individuals are to be considered as arising from their differing information bases. The term *objective degree of belief* may be used for such a notion.

The probability scale can be calibrated against a standard set of frequency-based chances. Arguments can again be produced as to why this form of probability obeys the usual rules of probability theory.

To be useful in individual applications, specific values have to be assigned to the probabilities and in many applications this is done by using a flat prior which is intended to represent an initial state of ignorance, leaving the final analysis to be essentially a summary of information provided by the data. Example 1.1, the normal mean, provides an instance where a very dispersed prior in the form of a normal distribution with very large variance v provides in the limit a Bayesian solution identical with the confidence interval form.

There are, however, some difficulties with this.

- Even for a scalar parameter θ the flat prior is not invariant under reparameterization. Thus if θ is uniform e^θ has an improper exponential distribution, which is far from flat.
- In a specific instance it may be hard to justify a distribution putting much more weight outside any finite interval than it does inside as a representation of ignorance or indifference.
- In multidimensional parameters the difficulties of specifying a suitable prior are much more acute.

For a scalar parameter the first point can be addressed by finding a form of the parameter closest to being a location parameter. One way of doing this will be discussed in Example 6.3.

The difficulty with multiparameter priors can be seen from the following example.

Example 5.5. *The noncentral chi-squared distribution.* Let $(Y_1, \ldots, Y_n)$ be independently normally distributed with unit variance and means $\mu_1, \ldots, \mu_n$ referring to independent situations and therefore with independent priors, assumed flat. Each μ_k^2 has posterior expectation $y_k^2 + 1$. Then if interest focuses on $\Delta^2 = \Sigma \mu_k^2$, it has posterior expectation $D^2 + n$. In fact its posterior distribution is noncentral chi-squared with n degrees of freedom and noncentrality $D^2 = \Sigma Y_k^2$. This implies that, for large n, Δ^2 is with high probability $D^2 + n + O_p(\sqrt{n})$. But this is absurd in that whatever the true value of Δ^2, the statistic D^2 is with high probability $\Delta^2 + n + O_p(\sqrt{n})$. A very flat prior in one dimension gives good results from almost all viewpoints, whereas a very flat prior and independence in many dimensions do not. This is called Stein's paradox or more accurately one of Stein's paradoxes.

If it were agreed that only the statistic D^2 and the parameter Δ^2 are relevant the problem could be collapsed into a one-dimensional one. Such a reduction is, in general, not available in multiparameter problems and even in this one a general Bayesian solution is not of this reduced form.

Quite generally a prior that gives results that are reasonable from various viewpoints for a single parameter will have unappealing features if applied independently to many parameters. The following example could be phrased more generally, for example for exponential family distributions, but is given now for binomial probabilities.

Example 5.6. *A set of binomial probabilities.* Let $\pi_1, \ldots, \pi_n$ be separate binomial probabilities of success, referring, for example, to properties of distinct parts of some random system. For example, success and failure may refer respectively to the functioning or failure of a component. Suppose that to estimate each probability m independent trials are made with r_k successes, trials for different events being independent. If each π_k is assumed to have a uniform prior on $(0, 1)$, then the posterior distribution of π_k is the beta distribution

$$\frac{\pi_k^{r_k}(1-\pi_k)^{m-r_k}}{B(r_k+1, m-r_k+1)}, \tag{5.7}$$

where the beta function in the denominator is a normalizing constant. It follows that the posterior mean of π_k is $(r_k+1)/(m+2)$. Now suppose that interest lies in some function of the π_k, such as in the reliability context $\psi_n = \Pi \pi_k$.

Because of the assumed independencies, the posterior distribution of ψ_n is derived from a product of beta-distributed random variables and hence is, for large m, close to a log normal distribution. Further, the mean of the posterior distribution is, by independence,

$$\Pi(r_k+1)/(m+2)^n \tag{5.8}$$

5.8 Impersonal degree of belief

and as $n \to \infty$ this, normalized by ψ_n, is

$$\prod \frac{1 + 1/(\pi_k m)}{1 + 2/m}. \tag{5.9}$$

Now especially if n is large compared with m this ratio is, in general, very different from 1. Indeed if all the π_k are small the ratio is greater than 1 and if all the π_k are near 1 the ratio is less than 1. This is clear on general grounds in that the probabilities encountered are systematically discrepant from the implications of the prior distribution.

This use of prior distributions to insert information additional to and distinct from that supplied by the data has to be sharply contrasted with an empirical Bayes approach in which the prior density is chosen to match the data and hence in effect to smooth the empirical distributions encountered. For this a simple approach is to assume a conjugate prior, in this case a beta density proportional to $\pi^{\lambda_1 - 1}(1 - \pi)^{\lambda_2 - 1}$ and having two unknown parameters. The marginal likelihood of the data, i.e., that obtained by integrating out π, can thus be obtained and the λs estimated by frequentist methods, such as those of Chapter 6. If errors in estimating the λs are ignored the application of Bayes' theorem to find the posterior distribution of any function of the π_k, such as ψ, raises no special problems. To make this into a fully Bayesian solution, it is necessary only to adopt a prior distribution on the λs; its form is unlikely to be critical.

The difficulties with flat and supposedly innocuous priors are most striking when the number of component parameters is large but are not confined to this situation.

Example 5.7. *Exponential regression.* Suppose that the exponential regression of Example 1.5 is rewritten in the form

$$E(Y_k) = \alpha + \beta \rho^{z_k}, \tag{5.10}$$

i.e., by writing $\rho = e^\gamma$, and suppose that it is known that $0 < \rho < 1$. Suppose further that α, β, ρ are given independent uniform prior densities, the last over $(0, 1)$ and that the unknown standard deviation σ has a prior proportional to $1/\sigma$; thus three of the parameters have improper priors to be regarded as limiting forms.

Suppose further that n independent observations are taken at values $z_k = z_0 + ak$, where $z_0, a > 0$. Then it can be shown that the posterior density of ρ tends to concentrate near 0 or 1, corresponding in effect to a model in which $E(Y)$ is constant.

5.9 Reference priors

While a notion of probability as measuring objective degree of belief has much intuitive appeal, it is unclear how to supply even approximate numerical values representing belief about sources of knowledge external to the data under analysis. More specifically to achieve goals similar to those addressed in frequentist theory it is required that the prior represents in some sense indifference or lack of information about relevant parameters. There are, however, considerable difficulties centring around the issue of defining a prior distribution representing a state of initial ignorance about a parameter. Even in the simplest situation of there being only two possible true distributions, it is not clear that saying that each has prior probability 1/2 is really a statement of no knowledge and the situation gets more severe when one contemplates infinite parameter spaces.

An alternative approach is to define a *reference prior* as a standard in which the contribution of the data to the resulting posterior is maximized without attempting to interpret the reference prior as such as a probability distribution.

A sketch of the steps needed to do this is as follows. First for any random variable W with known density $p(w)$ the *entropy* $\mathcal{H}(W)$ of a single observation on W is defined by

$$\mathcal{H}(W) = -E\{\log p(W)\} = -\int p(w) \log p(w) dw. \qquad (5.11)$$

This is best considered initially at least in the discrete form

$$-\Sigma p_k \log p_k, \qquad (5.12)$$

where the sum is over the points of support of W. The general idea is that observing a random variable whose value is essentially known is uninformative or unsurprising; the entropy is zero if there is unit probability at a single point, whereas if the distribution is widely dispersed over a large number of individually small probabilities the entropy is high. In the case of random variables, all discrete, representing Y and Θ we may compare the entropies of posterior and prior by

$$E_Y\{\mathcal{H}(\Theta \mid Y)\} - \mathcal{H}(\Theta), \qquad (5.13)$$

where the second term is the entropy of the prior distribution. In principle, the least informative prior is the one that maximizes this difference, i.e., which emphasizes the contribution of the data as contrasted with the prior.

Now suppose that the whole vector Y is independently replicated n times to produce a new random variable Y^{*n}. With the discrete parameter case contemplated, under mild conditions, and in particular if the parameter space is finite, the posterior distribution will become concentrated on a single point and the

5.9 Reference priors

first term of (5.13) will tend to zero and so the required prior has maximum entropy. That is, the prior maximizes (5.12) and for finite parameter spaces this attaches equal prior probability to all points.

There are two difficulties with this as a general argument. One is that if the parameter space is infinite $\mathcal{H}(\theta)$ can be made arbitrarily large. Also for continuous random variables the definition (5.11) is not invariant under monotonic transformations. To encapsulate the general idea of maximizing the contribution of the data in this more general setting requires a more elaborate formulation involving a careful specification of limiting operations which will not be given here. The argument does depend on limiting replication from Y to Y^{*n} as n increases; this is not the same as increasing the dimension of the observed random variable but rather is one of replication of the whole data.

Two explicit conclusions from this development are firstly that for a scale and location model, with location parameter μ and scale parameter σ, the reference prior has the improper form $d\mu d\sigma/\sigma$, i.e., it treats μ and $\log \sigma$ as having independent highly dispersed and effectively uniform priors. Secondly, for a single parameter θ the reference prior is essentially uniform for a transformation of the parameter that makes the model close to a location parameter. The calculation of this forms Example 6.3.

There are, however, difficulties with this approach, in particular in that the prior distribution for a particular parameter may well depend on which is the parameter of primary interest or even on the order in which a set of nuisance parameters is considered. Further, the simple dependence of the posterior on the data only via the likelihood regardless of the probability model is lost. If the prior is only a formal device and not to be interpreted as a probability, what interpretation is justified for the posterior as an adequate summary of information? In simple cases the reference prior may produce posterior distributions with especially good frequency properties but that is to show its use as a device for producing methods judged effective on other grounds, rather than as a conceptual tool.

Examples 5.5 and 5.6 show that flat priors in many dimensions may yield totally unsatisfactory posterior distributions. There is in any case a sharp contrast between such a use of relatively uninformative priors and the view that the objective of introducing the prior distribution is to inject additional information into the analysis. Many empirical applications of Bayesian techniques use proper prior distributions that appear flat relative to the information supplied by the data. If the parameter space to which the prior is attached is of relatively high dimension the results must to some extent be suspect. It is unclear at what point relatively flat priors that give good results from several viewpoints become unsatisfactory as the dimensionality of the parameter space increases.

5.10 Temporal coherency

The word *prior* can be taken in two rather different senses, one referring to time ordering and the other, in the present context, to information preceding, or at least initially distinct from, the data under immediate analysis. Essentially it is the second meaning that is the more appropriate here and the distinction bears on the possibility of data-dependent priors and of consistency across time.

Suppose that a hypothesis H is of interest and that it is proposed to obtain data to address H but that the data, to be denoted eventually by y, are not yet available. Then we may consider Bayes' theorem in the form

$$P^*(H \mid y) \propto P(H)P^*(y \mid H). \tag{5.14}$$

Here P^* refers to a probability of a hypothesized situation; we do not yet know y.

Later y becomes available and now

$$P(H \mid y) \propto P^*(H)P(y \mid H). \tag{5.15}$$

Now the roles of P and P^* have switched. The prior $P^*(H)$ is the prior that we would have if we did not know y, even though, in fact, we do know y. Temporal coherency is the assumption that, for example, $P(H) = P^*(H)$. In many situations this assumption is entirely reasonable. On the other hand it is not inevitable and there is nothing intrinsically inconsistent in changing prior assessments, in particular in the light of experience obtained either in the process of data collection or from the data themselves.

For example, a prior assessment is made in the design stage of a study, perhaps on the basis of previous similar studies or on the basis of well-established theory, say that a parameter ψ will lie in a certain range. Such assessments are a crucial part of study design, even if not formalized quantitatively in a prior distribution. The data are obtained and are in clear conflict with the prior assessment; for example ψ, expected to be negative, is pretty clearly positive.

There are essentially three possible explanations. The play of chance may have been unkind. The data may be contaminated. The prior may be based on a misconception. Suppose that reexamination of the theoretical arguments leading to the initial prior shows that there was a mistake either of formulation or even of calculation and that correction of that mistake leads to conformity with the data. Then it is virtually obligatory to allow the prior to change, the change having been driven by the data. There are, of course, dangers in such retrospective adjustment of priors. In many fields even initially very surprising effects can post hoc be made to seem plausible. There is a broadly analogous difficulty in frequentist theory. This is the selection of effects for significance testing after seeing the data.

Temporal coherency is a concern especially, although by no means only, in the context of studies continuing over a substantial period of time.

5.11 Degree of belief and frequency

In approaches to probability in which personalistic degree of belief is taken as the primary concept it is customary to relate that to frequency via a very long (infinite) sequence of exchangeable events. These can then be connected to a sequence of independent and identically distributed trials, each with the same probability p, and in which p itself has a probability distribution.

Suppose instead that $\mathcal{E}_1, \ldots, \mathcal{E}_n$ are n events, all judged by You to have approximately the same probability p and not to be strongly dependent. In fact they need not, in general, be related substantively. It follows from the Weak Law of Large Numbers obeyed by personalistic probability that Your belief that about a proportion p of the events are true has probability close to 1. You believe strongly that Your probabilities have a frequency interpretation.

This suggests that in order to elicit Your probability of an event $\mathcal{E}$ instead of contemplating hypothetical betting games You should try to find events $\mathcal{E}_1, \ldots, \mathcal{E}_n$ judged for good reason to have about the same probability as $\mathcal{E}$ and then find what proportion of this set is true. That would then indicate appropriate betting behaviour.

Of course, it is not suggested that this will be easily implemented. The argument is more one of fundamental principle concerning the primacy, even in a personalistically focused approach, of relating to the real world as contrasted with total concentration on internal consistency.

5.12 Statistical implementation of Bayesian analysis

To implement a Bayesian analysis, a prior distribution must be specified explicitly, i.e., in effect numerically. Of course, as with other aspects of formulation formal or informal sensitivity analysis will often be desirable.

There are at least four ways in which a prior may be obtained.

First there is the possibility of an implicitly empirical Bayes approach, where the word 'empirical' in this context implies some kind of frequency interpretation. Suppose that there is an unknown parameter referring to the outcome of a measuring process that is likely to be repeated in different circumstances and quite possibly by different investigators. Thus it might be a mean length in metres, or a mean duration in years. These parameters are likely to vary widely in different circumstances, even within one broad context, and hence to have a widely dispersed distribution. The use of a very dispersed prior distribution might then lead to a posterior distribution with a frequency interpretation over this ensemble of repetitions. While this interpretation has some appeal for simple parameters such as means or simple measures of dispersion, it is less appropriate for measures of dependence such as regression coefficients and contrasts of means. Also the independence assumptions involved in applying this to multiparameter problems are both implausible and potentially very misleading; see, for instance, Example 5.5.

The second approach is a well-specified empirical Bayes one. Suppose that a parameter with the same interpretation, for example as a difference between two treatments, is investigated in a large number of independent studies under somewhat different conditions. Then even for the parameter in, say, the first study there is some information in the other studies, how much depending on a ratio of components of variance.

Example 5.8. *Components of variance (ctd).* We start with a collapsed version of the model of Example 1.9 in which now there are random variables Y_k assumed to have the form

$$Y_k = \mu + \eta_k + \tilde{\epsilon}_k. \tag{5.16}$$

Here Y_k is the estimate of say the treatment contrast of interest obtained from the kth study and $\tilde{\epsilon}_k$ is the internal estimate of error in that study. That is, indefinite replication of the kth study would have produced the value $\mu + \eta_k$. The variation between studies is represented by the random variables η_k. We suppose that the ϵ and η are independently normally distributed with zero means and variances σ_ϵ^2 and σ_η^2. Note that in this version $\tilde{\epsilon}_k$ will depend on the internal replication within the study. While this formulation is an idealized version of a situation that arises frequently in applications, there are many aspects of it that would need critical appraisal in an application.

Now suppose that the number of studies is so large that μ and the components of variance may be regarded as known. If the mean of, say, the first group is the parameter of interest it has Y_1 as its direct estimate and also has the prior frequency distribution specified by a normal distribution of mean μ and variance σ_η^2 so that the posterior distribution has mean

$$\frac{Y_1/\sigma_\epsilon^2 + \mu/\sigma_\eta^2}{1/\sigma_\epsilon^2 + 1/\sigma_\eta^2}. \tag{5.17}$$

5.12 Statistical implementation of Bayesian analysis

Thus the direct estimate Y_1 is shrunk towards the general mean μ; this general effect does depend on the assumption that the relevant error distributions are not very long-tailed.

If now errors of estimation of μ and the variance components are to be included in the analysis there are two possibilities. The first possibility is to treat the estimation of μ and the variance components from a frequentist perspective. The other analysis, which may be called *fully Bayesian*, is to determine a prior density for μ and the variance components and to determine the posterior density of $\mu + \eta_1$ from first principles. In such a context the parameters μ and the variance components may be called *hyperparameters* and their prior distribution a *hyperprior*. Even so, within formal Bayesian theory they are simply unknowns to be treated like other unknowns.

Now in many contexts empirical replication of similar parameters is not available. The third approach follows the personalistic prescription. You elicit Your prior probabilities, for example via the kind of hypothetical betting games outlined in Section 5.7. For parameters taking continuous values a more realistic notion is to elicit the values of some parameters encapsulating Your prior distribution in some reasonably simple but realistic parametric form. Note that this distribution will typically be multidimensional and any specification of prior independence of component parameters may be critical.

In many, if not almost all, applications of statistical analysis the objective is to clarify what can reasonably be learned from data, at least in part as a base for public discussion within an organization or within a scientific field. For the approach to be applicable in such situations there needs both to be a self-denying ordinance that any prior probabilities are in some sense evidence-based, that base to be reported, and also for some approximate consensus on their values to be achieved.

What form might such evidence take? One possibility is that there are previous data which can reasonably be represented by a model in which some, at least, of the parameters, and preferably some of the parameters of interest, are in common with the current data. In this case, however, provided a consistent formulation is adopted, the same results would be achieved by a combined analysis of all data or, for that matter, by taking the current data as prior to the historical data. This is a consequence of the self-consistency of the laws of probability. Note that estimation of hyperparameters will be needed either by frequentist methods or via a specification of a prior for them. In this context the issue is essentially one of combining statistical evidence from different sources. One of the principles of such combination is that the mutual consistency of the information should be checked before merging. We return to this issue in the next section.

A second possibility is that the prior distribution is based on relatively informally recalled experience of a field, for example on data that have been seen only informally, or which are partly anecdotal, and which have not been carefully and explicitly analysed. An alternative is that it is based on provisional not-well-tested theory, which has not been introduced into the formulation of the probability model for the data. In the former case addition of an allowance for errors, biases and random errors in assessing the evidence would be sensible.

There is a broad distinction between two uses of such prior distributions. One is to deal with aspects intrinsic to the data under analysis and the other is to introduce external information. For example, in extreme cases the direct interpretation of the data may depend on aspects about which there is no information in the data. This covers such aspects as some kinds of systematic error, measurement errors in explanatory variables, the effect of unobserved explanatory variables, and so on. The choices are between considering such aspects solely qualitatively, or by formal sensitivity analysis, or via a prior distribution over parameters describing the unknown features. In the second situation there may be relevant partially formed theory or expert opinion external to the data. Dealing with expert opinion may raise delicate issues. Some research is designed to replace somewhat vaguely based expert opinion by explicit data-based conclusions and in that context it may be unwise to include expert opinion in the analysis, unless it is done in a way in which its impact can be disentangled from the contribution of the data. In some areas at least such expert opinion, even if vigorously expressed, may be fragile.

The next example discusses some aspects of dealing with systematic error.

Example 5.9. *Bias assessment.* An idealized version of the assessment of bias, or systematic error, is that $\bar{Y}$ is normally distributed with mean μ and known variance σ_0^2/n and that the parameter of interest ψ is $\mu + \gamma$, where γ is an unknown bias or systematic error that cannot be directly estimated from the data under analysis, so that assessment of its effect has to come from external sources.

The most satisfactory approach in many ways is a sensitivity analysis in which γ is assigned a grid of values in the plausible range and confidence intervals for ψ plotted against γ. Of course, in this case the intervals are a simple translation of those for μ but in more complex cases the relation might not be so straightforward and if there are several sources of bias a response

5.12 Statistical implementation of Bayesian analysis

surface would have to be described showing the dependence of the conclusions on components in question. It would then be necessary to establish the region in which the conclusions from $\bar{Y}$ would be seriously changed by the biases.

An alternative approach, formally possibly more attractive but probably in the end less insightful, is to elicit a prior distribution for γ, for example a normal distribution of mean ν and variance τ^2, based for instance on experience of similar studies for which the true outcome and hence the possible bias could be estimated. Then the posterior distribution is normal with mean a weighted average of $\bar{Y}$ and ν.

The emphasis on the second and third approaches to the choice of prior is on the refinement of analysis by inserting useful information, in the latter case information that is not directly assessable as a statistical frequency.

The final approach returns to the notion of a prior that will be generally accepted as a basis for assessing evidence from the data, i.e., which explicitly aims to insert as little new information as possible. For relatively simple problems we have already seen examples where limiting forms of prior reproduce approximately or exactly posterior intervals equivalent to confidence intervals. In complex problems with many parameters direct generalization of the simple priors produces analyses with manifestly unsatisfactory results; the notion of a flat or indifferent prior in a many-dimensional problem is untenable. The notion of a reference prior has been developed to achieve strong dependence on the data but can be complicated in implementation and the interpretation of the resulting posterior intervals, other than as approximate confidence intervals, is hard to fathom.

There is a conceptual conflict underlying this discussion. Conclusions expressed in terms of probability are on the face of it more powerful than those expressed indirectly via confidence intervals and p-values. Further, in principle at least, they allow the inclusion of a richer pool of information. But this latter information is typically more fragile or even nebulous as compared with that typically derived more directly from the data under analysis. The implication seems to be that conclusions derived from the frequentist approach are more immediately secure than those derived from most Bayesian analyses, except from those of a directly empirical Bayes kind. Any counter-advantages of Bayesian analyses come not from a tighter and more logically compelling theoretical underpinning but rather from the ability to quantify additional kinds of information.

Throughout the discussion so far it has been assumed that the model for the data, and which thus determines the likelihood, is securely based. The following section deals briefly with model uncertainty.

5.13 Model uncertainty

The very word *model* implies idealization of the real system and, except just possibly in the more esoteric parts of modern physics, as already noted in Section 3.1, it hardly makes sense to talk of a model being *true*.

One relatively simple approach is to test, formally or informally, for inconsistencies between the model and the data. Inconsistencies if found and statistically significant may nevertheless occasionally be judged unimportant for the interpretation of the data. More commonly inconsistencies if detected will require minor or major modification of the model. If no inconsistencies are uncovered it may be sufficient to regard the analysis, whether Bayesian or frequentist, as totally within the framework of the model.

Preferably, provisional conclusions having been drawn from an apparently well-fitting model, the question should be considered: have assumptions been made that might invalidate the conclusions? The answer will depend on how clear-cut the conclusions are and in many cases informal consideration of the question will suffice. In many cases analyses based on different models may be required; the new models may change the specification of secondary features of the model, for example by making different assumptions about the form of error distributions or dependence structures among errors. Often more seriously, the formulation of the primary questions may be changed, affecting the definition and interpretation of the parameter of interest ψ.

In frequentist discussions this process is informal; it may, for example, involve showing graphically or in tabular form the consequences of those different assumptions, combined with arguments about the general plausibility of different assumptions. In Bayesian discussions there is the possibility of the process of *Bayesian model averaging*. Ideally essentially the same subject-matter conclusions follow from the different models. Provided prior probabilities are specified for the different models and for the defining parameters within each model, then in principle a posterior probability for each model can be found and if appropriate a posterior distribution for a target variable found averaged across models, giving each its appropriate weight. The prior distributions for the parameters in each model must be proper.

Unfortunately for many interpretative purposes this procedure in inappropriate, quite apart from difficulties in specifying the priors. An exception is when empirical prediction of a new outcome variable, commonly defined for all models, is required; the notion of averaging several different predictions has some general appeal. The difficulty with interpretation is that seemingly similar parameters in different models rarely represent quite the same thing. Thus a regression coefficient of a response variable on a particular explanatory variable changes its interpretation if the outcome variable is transformed and more critically may depend drastically on which other explanatory variables are used. Further, if different reasonably well-fitting models give rather different conclusions it is important to know and report this.

5.14 Consistency of data and prior

Especially when a prior distribution represents potentially new important information, it is in principle desirable to examine the mutual consistency of information provided by the data and the prior. This will not be possible for aspects of the model for which there is little or no information in the data, but in general some comparison will be possible.

A serious discrepancy may mean that the prior is wrong, i.e., does not correspond to subject-matter reality, that the data are seriously biased or that the play of chance has been extreme.

Often comparison of the two sources of information can be informal but for a more formal approach it is necessary to find a distribution of observable quantities that is exactly or largely free of unknown parameters. Such a distribution is the marginal distribution of a statistic implied by likelihood and prior.

For example in the discussion of Example 1.1 about the normal mean, the marginal distribution implied by the likelihood and the normal prior is that

$$\frac{\bar{Y} - m}{\sqrt{(\sigma_0^2/n + v)}} \qquad (5.18)$$

has a standard normal distribution. That is, discrepancy is indicated whenever $\bar{Y}$ and the prior mean m are sufficiently far apart. Note that a very flat prior, i.e., one with extremely large v, will not be found discrepant.

A similar argument applies to other exponential family situations. Thus a binomial model with a beta conjugate prior implies a beta-binomial distribution for the number of successes, this distribution having known parameters and thus in principle allowing exact evaluation of the statistical significance of any departure.

5.15 Relevance of frequentist assessment

A key issue concerns the circumstances under which the frequentist interpretation of a p-value or confidence interval is relevant for a particular situation under study. In some rather general sense, following procedures that are in error relatively infrequently in the long run is some assurance for the particular case but one would like to go beyond that and be more specific.

Appropriate conditioning is one aspect already discussed. Another is the following. As with other measuring devices a p-value is calibrated in terms of the consequences of using it; also there is an implicit protocol for application that hinges on ensuring the relevance of the calibration procedure.

This protocol is essentially as follows. There is a question. A model is formulated. To help answer the question it may be that the hypothesis $\psi = \psi_0$ is considered. A test statistic is chosen. Data become available. The test statistic is calculated. In fact it will be relatively rare that this protocol is followed precisely in the form just set out.

It would be unusual and indeed unwise to start such an analysis without some preliminary checks of data completeness and quality. Corrections to the data would typically not affect the relevance of the protocol, but the preliminary study might suggest some modification of the proposed analysis. For example:

- some subsidiary aspects of the model might need amendment, for example it might be desirable to allow systematic changes in variance in a regression model;
- it might be desirable to change the precise formulation of the research question, for example by changing a specification of how $E(Y)$ depends on explanatory variables to one in which $E(\log Y)$ is considered instead;
- a large number of tests of distinct hypotheses might be done, all showing insignificant departures discarded, while reporting only those showing significant departure from the relevant null hypotheses;
- occasionally the whole focus of the investigation might change to the study of some unexpected aspect which nullified the original intention.

The third of these represents poor reporting practice but does correspond roughly to what sometimes happens in less blatant form.

It is difficult to specify criteria under which the departure from the protocol is so severe that the corresponding procedure is useless or misleading. Of the above instances, in the first two a standard analysis of the new model is likely to be reasonably satisfactory. In a qualitative way they correspond to fitting a broader model than the one originally contemplated and provided the fitting criterion is not, for example, chosen to maximize statistical significance, the results will be reasonably appropriate. That is not the case, however, for the last two possibilities.

Example 5.10. *Selective reporting.* Suppose that m independent sets of data are available each with its appropriate null hypothesis. Each is tested and p' is the

5.15 Relevance of frequentist assessment

smallest p-value achieved and H' the corresponding null hypothesis. Suppose that only H' and p' are reported. If say $m = 100$ it would be no surprise to find a p' as small as 0.01.

In this particular case the procedure followed is sufficiently clearly specified that a new and totally relevant protocol can be formulated. The test is based on the smallest of m independently distributed random variables, all with a uniform distribution under the overall null hypothesis of no effects. If the corresponding random variable is P', then

$$P(P' > x) = (1 - x)^m, \qquad (5.19)$$

because in order that $P' > x$ it is necessary and sufficient that all individual ps exceed x. Thus the significance level to be attached to p' under this scheme of investigation is

$$1 - (1 - p')^m \qquad (5.20)$$

and if p' is small and m not too large this will be close to mp'.

This procedure, named after Bonferroni, gives a quite widely useful way of adjusting simple p-values.

There is an extensive set of procedures, of which this is the simplest and most important, known under the name of *multiple comparison* methods. The name is, however, somewhat of a misnomer. Many investigations set out to answer several questions via one set of data and difficulties arise not so much from dealing with several questions, each answer with its measure of uncertainty, but rather from *selecting* one or a small number of questions on the basis of the apparent answer.

The corresponding Bayesian analysis requires a much more detailed specification and this point indeed illustrates one difference between the broad approaches. It would be naive to think that all problems deserve very detailed specification, even in those cases where it is possible in principle to set out such a formulation with any realism. Here for a Bayesian treatment it is necessary to specify both the prior probability that any particular set of null hypotheses is false and the prior distributions holding under the relevant alternative. Formally there may be no particular problem in doing this but for the prior to reflect genuine knowledge considerable detail would be involved. Given this formulation, the posterior distribution over the set of m hypotheses and corresponding alternatives is in principle determined. In particular the posterior probability of any particular hypothesis can be found.

Suppose now that only the set of data with largest apparent effect is considered. It would seem that if the prior distributions involve strong assumptions of independence among the sets of data, and of course that is not essential, then the information that the chosen set has the largest effect is irrelevant, the posterior distribution is unchanged, i.e., no direct allowance for selection is required.

A resolution of the apparent conflict with the frequentist discussion is, however, obtained if it is reasonable to argue that such a strategy of analysis is most likely to be used, if at all, when most of the individual null hypotheses are essentially correct. That is, with m hypotheses under examination the prior probability of any one being false may be approximately v_0/m, where v_0 may be treated as constant as m varies. Indeed v_0 might be approximately 1, so that the prior expectation is that one of the null hypotheses is false. The dependence on m is thereby restored.

An important issue here is that to the extent that the statistical analysis is concerned with the relation between data and a hypothesis about that data, it might seem that the relation should be unaffected by how the hypothesis came to be considered. Indeed a different investigator who had focused on the particular hypothesis H' from the start would be entitled to use p'. But if simple significance tests are to be used as an aid to interpretation and discovery in somewhat exploratory situations, it is clear that some such precaution as the use of (5.20) is essential to ensure relevance to the analysis as implemented and to avoid the occurrence of systematically wrong answers. In fact, more broadly, ingenious investigators often have little difficulty in producing convincing after-the-event explanations of surprising conclusions that were unanticipated beforehand but which retrospectively may even have high prior probability; see Section 5.10. Such ingenuity is certainly important but explanations produced by that route have, in the short term at least, different status from those put forward beforehand.

5.16 Sequential stopping

We have in much of the discussion set out a model in which the number n of primary observations is a constant not a random variable. While there are many contexts where this is appropriate, there are others where in some sense n could reasonably be represented as a random variable. Thus in many observational studies, even if there is a target number of observations, the number achieved may depart from this for a variety of reasons.

If we can reasonably add to the log likelihood a term for the probability density of the associated random variable N that is either known or, more realistically, depends on an unknown parameter γ, we may justify conditioning

5.16 Sequential stopping

on the observed sample size in the Bayesian formulation if the prior densities of γ and of θ are independent, or in the frequentist version if the parameters are variation-independent. This covers many applications and corresponds to current practice.

The situation is more difficult if the random variables are observed in sequence in such a way that, at least in a very formalized setting, when the first m observations, denoted collectively by $y^{(m)}$, are available, the probability that the $(m+1)$th observation is obtained is $p_{m+1}(y^{(m)})$ and that otherwise no more observations are obtained and $N = m$. In a formalized stopping rule the ps are typically 0 or 1, in that a decision about stopping is based solely on the current data. That is, it is assumed that in any discussion about whether to stop no additional information bearing on the parameter is used. The likelihood is essentially unchanged by the inclusion of such a purely data-dependent factor, so that, in particular, any Fisherian reduction of the data to sufficient statistics is unaffected after the inclusion of the realized sample size in the statistic; the values of intermediate data are then irrelevant.

In a Bayesian analysis, provided temporal coherency holds, i.e., that the prior does not change during the investigation, the stopping rule is irrelevant for analysis and the posterior density is computed from the likelihood achieved as if n were fixed. In a frequentist approach, however, this is not usually the case. In the simplest situation with a scalar parameter θ and a fixed-sample-size one-dimensional sufficient statistic s, the sufficient reduction is now only to the (2, 1) family (s, n) and it is not in general clear how to proceed.

In some circumstances it can be shown, or more often plausibly assumed as an approximation, that n is an ancillary statistic. That is to say, knowledge of n on its own would give little or no information about the parameter of interest. Then it will be reasonable to condition on the value of sample size and to consider the conditional distribution of S for fixed sample size, i.e., to ignore the particular procedure used to determine the sample size.

Example 5.11. *Precision-based choice of sample size.* Suppose that the mean of a normal distribution, or more generally a contrast of several means, is being estimated and that at each stage of the analysis the standard error of the mean or contrast is found by the usual procedure. Suppose that the decision to stop collecting data is based on these indicators of precision, not of the estimates of primary concern, the mean or contrast of means. Then treating sample size as fixed defines a reasonable calibration of the p-value and confidence limits.

The most extreme departure from treating sample size as fixed arises when the statistic s, typically a sum or mean value is fixed. Then it is the inverse distribution, i.e., of N given $S = s$, that is relevant. It can be shown in at least some cases the difference from the analysis treating sample size as fixed is small.

Example 5.12. *Sampling the Poisson process.* In any inference about the Poisson process of rate ρ the sufficient statistic is (N, S), the number of points observed and the total time of observation. If S is fixed, N has a Poisson distribution of mean ρs, whereas if N is fixed S has the gamma density of mean n/ρ, namely

$$\rho(\rho s)^{(n-1)} e^{-\rho s}/(n-1)!. \tag{5.21}$$

To test the null hypothesis $\rho = \rho_0$ looking for one-sided departures in the direction $\rho > \rho_0$ there are two p-values corresponding to the two modes of sampling, namely

$$\Sigma_{r=n}^{\infty} e^{-\rho_0 s} (\rho_0 s)^r / r! \tag{5.22}$$

and

$$\int_0^s \{e^{-\rho_0 x} (\rho_0 x)^{n-1} \rho_0 / (n-1)!\} dx. \tag{5.23}$$

Repeated integration by parts shows that the integral is equal to (5.22) as is clear from a direct probabilistic argument.

Very similar results hold for the binomial and normal distributions, the latter involving the inverse Gaussian distribution.

The primary situations where there is an appreciable effect of the stopping rule on the analysis are those where the emphasis is strongly on the testing of a particular null hypothesis, with attention focused strongly or even exclusively on the resulting p-value. This is seen most strongly by the procedure that stops when and only when a preassigned level of significance is reached in formal testing of a given null hypothesis. There is an extensive literature on the sequential analysis of such situations.

It is here that one of the strongest contrasts arises between Bayesian and frequentist formulations. In the Bayesian approach, provided that the prior and model remain fixed throughout the investigation, the final inference, in particular the posterior probability of the null hypothesis, can depend on the data only through the likelihood function at the terminal stage. In the frequentist approach, the final interpretation involves the data directly not only via that same likelihood function but also on the stopping criterion used, and so in particular on the information that stopping did not take place earlier. It should involve the specific earlier data only if issues of model adequacy are involved; for example it might be suspected that the effect of an explanatory variable had been changing in time.

5.17 A simple classification problem

A very idealized problem that illuminates some aspects of the different approaches concerns a classification problem involving the allocation of a new individual to one of two groups, G_0 and G_1. Suppose that this is to be done on the basis of an observational vector y, which for group G_k has known density $f_k(y)$ for $k = 0, 1$.

For a Bayesian analysis suppose that the prior probabilities are π_k, with $\pi_0 + \pi_1 = 1$. Then the posterior probabilities satisfy

$$\frac{P(G_1 \mid Y = y)}{P(G_0 \mid Y = y)} = \frac{\pi_1 f_1(y)}{\pi_0 f_0(y)}. \quad (5.24)$$

One possibility is to classify an individual into the group with the higher posterior probability, another is to record the posterior probability, or more simply the posterior odds, as a succinct measure of evidence corresponding to each particular y.

The issue of a unique assignment becomes a well-formulated decision problem if relative values can be assigned to w_{01} and w_{10}, the losses of utility following from classifying an individual from group G_0 into group G_1 and vice versa. A rule minimizing expected long-run loss of utility is to classify to group G_1 if and only if

$$\frac{\pi_1 f_1(y)}{\pi_0 f_0(y)} \geq \frac{w_{01}}{w_{10}}. \quad (5.25)$$

The decision can be made arbitrarily in the case of exact equality.

The data enter only via the likelihood ratio $f_1(y)/f_0(y)$. One non-Bayesian resolution of the problem would be to use just the likelihood ratio itself as a summary measure of evidence. It has some direct intuitive appeal and would allow any individual with a prior probability to make an immediate assessment.

We may, however, approach the problem also more in the spirit of a direct frequency interpretation involving hypothetical error rates. For this, derivation of a sufficiency reduction is the first step. Write the densities in the form

$$f_0(y) \exp[\log\{f_1(y)/f_0(y)\}\psi - k(\psi)], \quad (5.26)$$

where the parameter space for the parameter of interest ψ consists of the two points 0 and 1. The sufficient statistic is thus the likelihood ratio.

Now consider testing the null hypothesis that the observation y comes from the distribution $f_0(y)$, the test to be sensitive against the alternative distribution $f_1(y)$. This is a radical reformulation of the problem in that it treats the two distributions asymmetrically putting emphasis in particular on $f_0(y)$ as a null hypothesis. The initial formulation as a classification problem treats both distributions essentially symmetrically. In the significance testing formulation we use the likelihood ratio as the test statistic, in virtue of the sufficiency reduction and calculate as p-value

$$\int_{\mathcal{L}_{\text{obs}}} f_0(y) dy, \tag{5.27}$$

where $\mathcal{L}_{\text{obs}}$ is the set of points with a likelihood ratio as great or greater than that observed. In the Neyman–Pearson theory of testing hypotheses this is indirectly the so-called Fundamental Lemma.

A formally different approach to this issue forms the starting point to the Neyman–Pearson theory of testing hypotheses formalized directly in the Fundamental Lemma. This states, in the terminology used here, that of all test statistics that might be used, the probability of exceeding a preassigned p-value is maximized under the alternative $f_1(y)$ by basing the test on the likelihood ratio. Proofs are given in most books on statistical theory and will be omitted.

Symmetry between the treatment of the two distributions is restored by in effect using the relation between tests and confidence intervals. That is, we test for consistency with $f_0(y)$ against alternative $f_1(y)$ and then test again interchanging the roles of the two distributions. The conclusion is then in the general form that the data are consistent with both specified distributions, with one and not the other, or with neither.

Of course, this is using the data for interpretative purposes not for the classification aspect specified in the initial part of this discussion. For the latter purpose summarization via the likelihood ratio is simple and direct.

Example 5.13. *Multivariate normal distributions.* A simple example illustrating the above discussion is to suppose that the two distributions are multivariate normal distributions with means μ_0, μ_1 and known covariance matrix Σ_0. We may, without loss of generality, take $\mu_0 = 0$. The log likelihood ratio is then

$$-(y-\mu_1)^T \Sigma_0^{-1}(y-\mu_1)/2 + (y^T \Sigma_0^{-1} y)/2 = y^T \Sigma_0^{-1} \mu_1 - \mu_1^T \Sigma_0^{-1} \mu_1/2. \tag{5.28}$$

The observations thus enter via the linear discriminant function $y^T \Sigma_0^{-1} \mu_1$. The various more detailed allocation rules and procedures developed above follow immediately.

The geometric interpretation of the result is best seen by a preliminary linear transformation of y so that it has the identity matrix as covariance matrix and then the linear discriminant is obtained by orthogonal projection of y onto the vector difference of means. In terms of the original ellipsoid the discriminant is in the direction conjugate to the vector joining the means; see Figure 5.1.

If the multivariate normal distributions have unequal covariance matrices the discriminant function becomes quadratic in y having the interpretation that an

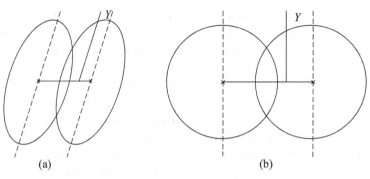

Figure 5.1. (a) The ellipses are contours of constant density for two bivariate normal distributions with the same covariance matrix. The dashed lines show the direction conjugate to the line joining means. Projection of the observed random variable Y parallel to conjugate direction determines linear discriminant. (b) After linear transformation component variables are independent with the same variance, ellipses become circles and the discriminant corresponds to orthogonal projection onto the line joining centres.

observation very distant from both vector means is more likely to come from the distribution with larger covariance matrix contribution in the direction of departure. An extreme case is when the two distributions have the same vector mean but different covariance matrices.

Notes 5

Section 5.3. For the contradictions that may follow from manipulating fiducial probabilities like ordinary probabilities, see Lindley (1958). For the exchange paradox, see Pawitan (2000) and for a more detailed discussion Christensen and Utts (1992).

Section 5.5. The discussion of rain in Gothenburg is an attempt to explain R. A. Fisher's notion of probability.

Section 5.6. Robinson (1979) discusses the difficulties with the notion of recognizable subsets. Kalbfleisch (1975) separates the two different reasons for conditioning. For a distinctive account of these issues, see Sundberg (2003).

Section 5.7. For more references on the history of these ideas, see Appendix A. See Garthwaite *et al.* (2005) for a careful review of procedures for eliciting expert opinion in probabilistic form. See also Note 4.2. Walley (1991) gives a thorough account of an approach based on the notion that You can give only upper and lower bounds to Your probabilities.

Section 5.8. Again see Appendix A for further references. Stein (1959) gives Example 5.5 as raising a difficulty with fiducial probability; his arguments are just as cogent in the present context. Mitchell (1967) discusses Example 5.7.

Section 5.9. Bernardo (1979) introduces reference priors and gives (Bernardo, 2005) a careful account of their definition, calculation and interpretation. Jaynes (1976) discusses finite parameter spaces. Berger (2004) defends the impersonal approach from a personalistic perspective. Kass and Wasserman (1996) give a comprehensive review of formal rules for selecting prior distributions. Lindley (1956) develops the notion of information in a distribution and its relation to Shannon information and to entropy.

Section 5.11. The argument in this section is variously regarded as wrong, as correct but irrelevant, and as correct and important to a degree of belief concept of probability.

Section 5.12. Greenland (2005) gives a searching discussion of systematic errors in epidemiological studies, arguing from a Bayesian viewpoint.

Section 5.13. Copas and Eguchi (2005) discuss the effects of local model uncertainty. A key issue in such discussions is the extent to which apparently similar parameters defined in different models retain their subject-matter interpretation.

Section 5.15. Hochberg and Tamhane (1987) give a systematic account of multiple comparison methods. For work emphasizing the notion of false discovery rate, see Storey (2002).

Section 5.16. A treatment of sequential stopping by Wald (1947) strongly based on the use of the likelihood ratio between two base hypotheses led to very extensive development. It was motivated by applications to industrial inspection. More recent developments have focused on applications to clinical trials (Whitehead, 1997). Anscombe (1949, 1953) puts the emphasis on estimation and shows that to a first approximation special methods of analysis are not needed, even from a frequentist viewpoint. See also Barndorff-Nielsen and Cox (1984). Vaagerö and Sundberg (1999) discuss a data-dependent design, the up-and-down method of estimating a binary dose–response curve, in which it is the design rather than the stopping rule that distorts the likelihood-based estimation of the slope of the curve, tending to produce estimates of slope that are too steep. It is not clear whether this is a case of failure of standard asymptotics or of very slow convergence to the asymptotic behaviour.

Section 5.17. The discrimination problem has been tackled in the literature from many points of view. The original derivation of the linear discriminant function (Fisher, 1936) is as that linear function of the data, normalized to have fixed variance, that maximized the difference between the means. As such it has probably been more used as a tool of analysis than for discrimination. Later Fisher (1940) notes that by scoring the two groups say 0 and 1 and making a least squares analysis of that score on the measured variables as fixed, not only was the linear discriminant function recovered as the regression equation, but the significance of coefficients and groups of coefficients could be tested by the standard normal-theory tests as if the (0, 1) score had a normal distribution and the measured variables were fixed.

To interpret Figure 5.1, in either of the defining ellipses draw chords parallel to the line joining the two means. The locus of midpoints of these chords is a line giving the conjugate direction.

The order in probability notation $O_p(1/\sqrt{n})$ is used extensively in Chapter 6; see Note 6.2.

6

Asymptotic theory

Summary. Approximate forms of inference based on local approximations of the log likelihood in the neighbourhood of its maximum are discussed. An initial discussion of the exact properties of log likelihood derivatives includes a definition of Fisher information. Then the main properties of maximum likelihood estimates and related procedures are developed for a one-dimensional parameter. A notation is used to allow fairly direct generalization to vector parameters and to situations with nuisance parameters. Finally numerical methods and some other issues are discussed in outline.

6.1 General remarks

The previous discussion yields formally exact frequentist solutions to a number of important problems, in particular concerning the normal-theory linear model and various problems to do with Poisson, binomial, exponential and other exponential family distributions. Of course the solutions are formal in the sense that they presuppose a specification which is at best a good approximation and which may in fact be inadequate. Bayesian solutions are in principle always available, once the full specification of the model and prior distribution are established.

There remain, however, many situations for which the exact frequentist development does not work; these include nonstandard questions about simple situations and many models where more complicated formulations are unavoidable for any sort of realism. These issues are addressed by asymptotic analysis. That is, approximations are derived on the basis that the amount of information is large, errors of estimation are small, nonlinear relations are locally linear and a central limit effect operates to induce approximate normality of log likelihood derivatives. The corresponding Bayesian results are concerned with finding approximations to the integrals involved in calculating a posterior density.

6.2 Scalar parameter

6.2.1 Exact results

It is simplest to begin with a one-dimensional unknown parameter θ and log likelihood $l(\theta; Y)$ considered as a function of the vector random variable Y, with gradient

$$U = U(\theta) = U(\theta; Y) = \partial l(\theta; Y)/\partial \theta. \tag{6.1}$$

This will be called the *score* or more fully the *score function*. Now, provided the normalizing condition

$$\int f_Y(y; \theta) dy = 1 \tag{6.2}$$

can be differentiated under the integral sign with respect to θ, we have that

$$\int U(\theta; y) f_Y(y; \theta) dy = 0. \tag{6.3}$$

That is,

$$E\{U(\theta; Y); \theta\} = 0, \tag{6.4}$$

where the expectation and differentiation take place at the same value of θ. If it is legitimate to differentiate also (6.3) under the integral sign it follows that

$$\int \{\partial^2 l(\theta; Y)/\partial \theta^2 + U^2(\theta; Y)\} f_Y(y; \theta) dy = 0. \tag{6.5}$$

Therefore if we define

$$i(\theta) = E\{-\partial^2 l(\theta; Y)/\partial \theta^2; \theta\}, \tag{6.6}$$

then

$$\text{var}\{U(\theta; Y); \theta\} = i(\theta). \tag{6.7}$$

The function $i(\theta)$ is called the *expected information* or sometimes the *Fisher information* about θ. Note that it and the gradient U are calculated from the vector Y representing the full vector of observed random variables. Essentially $i(\theta)$ measures the expected curvature of the log likelihood and the greater the curvature the sharper the inference.

Example 6.1. *Location model (ctd).* If Y is a single random variable from the density $g(y - \theta)$, the score is

$$\frac{dg(y-\theta)/d\theta}{g(y-\theta)} \tag{6.8}$$

and the information is

$$\int [\{dg(y)/dy\}^2/g(y)]dy, \tag{6.9}$$

sometimes called the *intrinsic accuracy* of the family $g(.)$. That the information is independent of θ is one facet of the variation inherent in the model being independent of θ. For a normal distribution of unknown mean and known variance σ_0^2 the intrinsic accuracy is $1/\sigma_0^2$.

Example 6.2. *Exponential family.* Suppose that Y corresponds to a single observation from the one-parameter exponential family distribution with canonical parameter θ, the distribution thus being written in the form

$$m(y)\exp\{\theta s - k(\theta)\}. \tag{6.10}$$

Then the score function is $s - dk(\theta)/d\theta = s - \eta$, where η is the mean parameter, and the information is $d^2k(\theta)/d\theta^2$, which is equal to $\text{var}(S;\theta)$. The information is independent of θ if and only if $k(\theta)$ is quadratic. The distribution is then normal, the only exponential family distribution which is a special case of a location model in the sense of Example 6.1.

By an extension of the same argument, in which the new density $f_Y(y;\theta+\delta)$ is expanded in a series in δ, we have

$$E\{U(\theta);\theta+\delta\} = i(\theta)\delta + O(\delta^2). \tag{6.11}$$

This shows two roles for the expected information.

Discussion of the regularity conditions involved in these and subsequent arguments is deferred to Chapter 7.

6.2.2 Parameter transformation

In many applications the appropriate form for the parameter θ is clear from a subject-matter interpretation, for example as a mean or as a regression coefficient. From a more formal viewpoint, however, the whole specification is unchanged by an arbitrary smooth $(1,1)$ transformation from θ to, say, $\phi = \phi(\theta)$. When such transformations are under study, we denote the relevant score and information functions by respectively $U^\Theta(\theta;Y), i^\Theta(\theta)$ and $U^\Phi(\phi;Y), i^\Phi(\phi)$.

6.2 Scalar parameter

Directly from the definition

$$U^\Phi\{\phi(\theta); Y\} = U^\Theta(\theta; Y) d\theta/d\phi \qquad (6.12)$$

and, because the information is the variance of the score,

$$i^\Phi\{\phi(\theta)\} = i^\Theta(\theta)(d\theta/d\phi)^2. \qquad (6.13)$$

Example 6.3. *Transformation to near location form.* If we choose the function $\phi(\theta)$ so that

$$d\phi(\theta)/d\theta = \sqrt{i^\Theta(\theta)}, \qquad (6.14)$$

i.e., so that

$$\phi(\theta) = \int_a^\theta \sqrt{i^\Theta(x)} dx, \qquad (6.15)$$

then the information is constant in the ϕ-parameterization. In a sense, the nearest we can make the system to a pure location model is by taking this new parameterization. The argument is similar to but different from that sometimes used in applications, of transforming the observed random variable aiming to make its variance not depend on its mean.

A prior density in which the parameter ϕ is uniform is called a *Jeffreys prior*.

For one observation from the exponential family model of Example 6.2 the information about the canonical parameter θ is $i^\Theta(\theta) = k''(\theta)$. The information about the mean parameter $\eta = k'(\theta)$ is thus

$$i^\eta(\theta) = i^\Theta(\theta)/(d\eta/d\theta)^2 = 1/k''(\theta), \qquad (6.16)$$

the answer being expressed as a function of η. This is also the reciprocal of the variance of the canonical statistic.

In the special case of a Poisson distribution of mean μ, written in the form

$$(1/y!) \exp(y \log \mu - \mu), \qquad (6.17)$$

the canonical parameter is $\theta = \log \mu$, so that $k(\theta) = e^\theta$. Thus, on differentiating twice, the information about the canonical parameter is $e^\theta = \mu$, whereas the information about $\eta = \mu$ is $1/\mu$. Further, by the argument of Example 6.3, the information is constant on taking as new parameter

$$\phi(\theta) = \int_a^\theta e^{x/2} dx \qquad (6.18)$$

and, omitting largely irrelevant constants, this suggests transformation to $\phi = e^{\theta/2}$, i.e., to $\phi = \sqrt{\mu}$. Roughly speaking, the transformation extends the part of the parameter space near the origin and contracts the region of very large values.

6.2.3 Asymptotic formulation

The results so far involve no mathematical approximation. Suppose now that there is a quantity n which is frequently a sample size or other measure of the amount of data, but which more generally is an index of the precision of the resulting estimates, such as a reciprocal variance of a measurement procedure. The following mathematical device is now used to generate approximations.

We denote the mean expected information per unit information, $i(\theta)/n$, by $\bar{\imath}(\theta)$ and assume that as $n \to \infty$ it has a nonzero limit. Further, we assume that $U(\theta)/\sqrt{n}$ converges in distribution to a normal distribution of zero mean and variance $\bar{\imath}(\theta)$. Note that these are in fact the exact mean and limiting variance.

Results obtained by applying the limit laws of probability theory in this context are to be regarded as approximations hopefully applying to the data and model under analysis. The limiting operation $n \to \infty$ is a fiction used to derive these approximations whose adequacy in principle always needs consideration. There is never any real sense in which $n \to \infty$; the role of asymptotic analysis here is no different from that in any other area of applied mathematics.

We take the maximum likelihood estimate to be defined by the equation

$$U(\hat{\theta}; Y) = 0, \qquad (6.19)$$

taking the solution giving the largest log likelihood should there be more than one solution. Then the most delicate part of the following argument is to show that the maximum likelihood estimate is very likely to be close to the true value. We argue somewhat informally as follows.

By assumption then, as $n \to \infty$, $i(\theta)/n \to \bar{\imath}(\theta) > 0$, and moreover $l(\theta)$ has the qualitative behaviour of the sum of n independent random variables, i.e., is $O_p(n)$ with fluctuations that are $O_p(\sqrt{n})$. Finally it is assumed that $U(\theta; Y)$ is asymptotically normal with zero mean and variance $i(\theta)$. We call these conditions *standard n asymptotics*. Note that they apply to time series and spatial models with short-range dependence. For processes with long-range dependence different powers of n would be required in the normalization. For a review of the o, O, o_p, O_p notation, see Note 6.2.

It is convenient from now on to denote the notional true value of the parameter by θ^*, leaving θ as the argument of the log likelihood function. First, note that by Jensen's inequality

$$E\{l(\theta; Y); \theta^*\} \qquad (6.20)$$

has its maximum at $\theta = \theta^*$. That is, subject to smoothness, $E\{l(\theta; Y)\}$ as a function of θ has its maximum at the true value θ^*, takes a value that is $O(n)$ and has a second derivative at the maximum that also is $O(n)$. Thus unless $|\theta - \theta^*| = O(1/\sqrt{n})$ the expected log likelihood is small compared with its

6.2 Scalar parameter

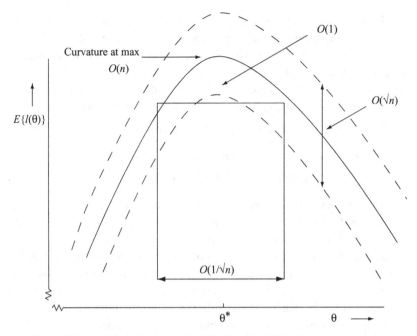

Figure 6.1. Behaviour of expected log likelihood (*solid line*) near the true value within the band of width $O(\sqrt{n})$ around the expected value. Asymptotic behaviour near maximum as indicated.

maximum value, and the likelihood itself is exponentially small compared with its maximum value.

Next note that the observed log likelihood will lie within $O_p(\sqrt{n})$ of its expectation and because of the smoothness of the observed log likelihood this statement applies to the whole function over a range within $O(1/\sqrt{n})$ of θ^*, so that $\hat{\theta}$ lies within $O_p(1/\sqrt{n})$ of θ^*. See Figure 6.1.

The argument does not exclude the occurrence of other local maxima to $l(\theta)$ but typically the log likelihood evaluated there will be very much less than $l(\hat{\theta})$, the overall maximum. Of course, the possibility of multiple maxima may well be a considerable concern in implementing iterative schemes for finding $\hat{\theta}$.

In the light of the above discussion, we expand $l(\theta)$ about θ^* to quadratic terms to obtain

$$l(\theta) - l(\theta^*) = U(\theta^*)(\theta - \theta^*) - (\theta - \theta^*)^2 j(\theta^*)/2, \quad (6.21)$$

where

$$j(\theta^*) = -\partial^2 l(\theta^*)/\partial \theta^2 \quad (6.22)$$

is called the *observed information function at* θ^* in the light of its connection with $i(\theta^*)$, the expected information. In the range of interest within $O_p(1/\sqrt{n})$ of θ^* the next term in the expansion will be negligible under mild restrictions on the third derivative of $l(\theta)$.

It follows, on differentiating (6.21), then obtaining the corresponding expansion for $U(\theta)$, and finally because $U(\hat{\theta}) = 0$, that approximately

$$\hat{\theta} - \theta^* = j^{-1}(\theta^*)U(\theta^*), \qquad (6.23)$$

which to achieve appropriate normalization is better written

$$(\hat{\theta} - \theta^*)\sqrt{n} = \{nj^{-1}(\theta^*)\}\{U(\theta^*)/\sqrt{n}\}. \qquad (6.24)$$

Now $j(\theta^*)/n$ converges in probability to its expectation $i(\theta^*)/n$ which itself has been assumed to converge to $\bar{\iota}(\theta^*)$, the mean expected information per unit n.

We appeal to the result that if the sequence of random variables Z_n converges in distribution to a normal distribution of zero mean and variance σ^2 and if the sequence A_n converges in probability to the nonzero constant a, then the sequence $A_n Z_n$ converges in distribution to a normal distribution of zero mean and variance $a^2\sigma^2$. That is, variation in the A_n can be ignored. This can be proved by examining the distribution function of the product.

Now, subject to the continuity of $i(\theta)$, the following ratios all tend in probability to 1 as $n \to \infty$:

$$i(\hat{\theta})/i(\theta^*); \quad j(\theta^*)/i(\theta^*); \quad j(\hat{\theta})/i(\theta^*). \qquad (6.25)$$

This means that in the result for the asymptotic distribution of the maximum likelihood estimate the asymptotic variance can be written in many asymptotically equivalent forms. In some ways the most useful of these employs the third form involving what we now denote by $\hat{j}$ and have called simply the *observed information*, namely

$$\hat{j} = j(\hat{\theta}). \qquad (6.26)$$

It follows from (6.24) and (6.25) that, the maximum likelihood estimate is asymptotically normally distributed with mean θ^* and variance that can be written in a number of forms, in particular $\hat{j}^{-1}$ or $i^{-1}(\theta^*)$.

Suppose that we make a (1, 1) transformation of the parameter from θ to $\phi(\theta)$. The likelihood at any point is unchanged by the labelling of the parameters so that the maximum likelihood estimates correspond exactly, $\hat{\phi} = \phi(\hat{\theta})$. Further, the information in the ϕ-parameterization that gives the asymptotic variance of $\hat{\phi}$ is related to that in the θ-parameterization by (6.13) so that the asymptotic variances are related by

$$\text{var}(\hat{\phi}) = \text{var}(\hat{\theta})(d\phi/d\theta)^2. \qquad (6.27)$$

6.2 Scalar parameter

This relation follows also directly by local linearization of $\phi(\theta)$, the so-called delta method. Note that also, by the local linearity, asymptotic normality applies to both $\hat{\theta}$ and $\hat{\phi}$, but that in any particular application one version might be appreciably closer to normality than another.

Thus we may test the null hypothesis $\theta = \theta_0$ by the test statistic

$$(\hat{\theta} - \theta_0)/\sqrt{i^{-1}(\theta_0)}, \tag{6.28}$$

or one of its equivalents, and confidence intervals can be found either from the set of values consistent with the data at some specified level or more conveniently from the pivot

$$(\theta - \hat{\theta})/\sqrt{\hat{j}^{-1}}. \tag{6.29}$$

6.2.4 Asymptotically equivalent statistics

Particularly in preparation for the discussion with multidimensional parameters it is helpful to give some further discussion and comparison formulated primarily in terms of two-sided tests of a null hypothesis $\theta = \theta_0$ using quadratic test statistics.

That is, instead of the normally distributed statistic (6.28) we use its square, written, somewhat eccentrically, as

$$W_E = (\hat{\theta} - \theta_0) i(\theta_0)(\hat{\theta} - \theta_0). \tag{6.30}$$

Because under the null hypothesis this is the square of a standardized normal random variable, W_E has under that hypothesis a chi-squared distribution with one degree of freedom.

In the form shown in Figure 6.2, the null hypothesis is tested via the squared horizontal distance between the maximum likelihood estimate and the null value, appropriately standardized.

An alternative procedure is to examine the vertical distance, i.e., to see how much larger a log likelihood is achieved at the maximum than at the null hypothesis. We may do this via the likelihood ratio statistic

$$W_L = 2\{l(\hat{\theta}) - l(\theta_0)\}. \tag{6.31}$$

Appeal to the expansion (6.21) shows that W_L and W_E are equal to the first order of asymptotic theory, in fact that

$$W_E - W_L = O_p(1/\sqrt{n}). \tag{6.32}$$

Thus under the null hypothesis W_L has a chi-squared distribution with one degree of freedom and by the relation between significance tests and confidence

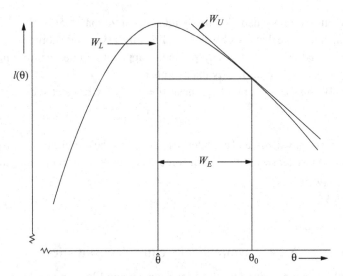

Figure 6.2. Three asymptotically equivalent ways, all based on the log likelihood function of testing null hypothesis $\theta = \theta_0$: W_E, horizontal distance; W_L vertical distance; W_U slope at null point.

regions a confidence set for θ is given by taking all those values of θ for which $l(\theta)$ is sufficiently close to $l(\hat{\theta})$. That is,

$$\{\theta : l(\hat{\theta}) - l(\theta) \leq k^{*2}_{1;c}/2\}, \tag{6.33}$$

where $k^{*2}_{1;c}$ is the upper c point of the chi-squared distribution with one degree of freedom, forms a $1 - c$ level confidence region for θ. See Figure 6.2. Note that unless the likelihood is symmetric around $\hat{\theta}$ the region will not be symmetric. To obtain a one-sided version of W_L more directly comparable to $(\hat{\theta} - \theta_0)\sqrt{\hat{\jmath}}$ we may define the signed likelihood ratio statistic as

$$\text{sgn}(\hat{\theta} - \theta_0)\sqrt{W_L}, \tag{6.34}$$

treating this as having approximately a standard normal distribution when $\theta = \theta_0$.

There is a third possibility, sometimes useful in more complicated problems, namely to use the gradient of the log likelihood at θ_0; if this is too large it suggests that the data can be explained much better by changing θ. See again Figure 6.2.

To implement this idea we use the statistic

$$T_U = U(\theta_0)/\sqrt{i(\theta_0)}, \tag{6.35}$$

6.2 Scalar parameter

i.e., the slope divided by its standard error. Note that for direct comparability with the other two one-sided statistics a change of sign is needed.

In the quadratic version we take

$$W_U = U(\theta_0; Y) i^{-1}(\theta_0) U(\theta_0; Y). \tag{6.36}$$

It follows directly from the previous discussion that this is to the first order equal to both W_E and W_L.

Note that W_E and W_L would be identical were the log likelihood exactly quadratic. Then they would be identical also to W_U were observed rather than expected information used in the definitions. Under the standard conditions for asymptotic theory this quadratic behaviour is approached locally and asymptotically. In applications the three statistics are very often essentially equal for practical purposes; any serious discrepancy between them indicates that the standard theory is inadequate in at least some respects. We leave to Section 6.6 the discussion of what to do then and the broader issue of the relative advantages of the forms we have given here.

6.2.5 Optimality considerations

One approach to the discussion of the optimality of these procedures is by detailed analysis of the properties of the resulting estimates and tests. A more direct argument is as follows. In the previous discussion we expanded the log likelihood around the true value. We now expand instead around the maximum likelihood estimate $\hat{\theta}$ to give for θ sufficiently close to $\hat{\theta}$ that

$$l(\theta) = l(\hat{\theta}) - (\theta - \hat{\theta})^2 \hat{j}/2 + O(1/\sqrt{n}), \tag{6.37}$$

or, on exponentiating, that the likelihood has the form

$$m(y) \exp\{-(\theta - \hat{\theta})^2 \hat{j}/2\} \{1 + O(1/\sqrt{n})\}, \tag{6.38}$$

where $m(y)$ is a normalizing function. This gives an approximate version of the Fisherian reduction and shows that to the first order of asymptotic theory the data can be divided into $(\hat{\theta}, \hat{j})$, which summarize what can be learned given the model, and the data conditionally on $(\hat{\theta}, \hat{j})$, which provide information about model adequacy. That is, the maximum likelihood estimate and the observed information are approximately sufficient statistics. Note that $\hat{\theta}$ could be replaced by any estimate differing from it by $O_p(n^{-1})$ and there are many such!

6.2.6 Bayesian analysis

In principle once the Bayesian formulation has been established, the calculation of the posterior distribution involves only integration and so the role of asymptotic theory, at least in one dimension, is restricted to establishing the general form of the answers to be expected and often to simplifying the calculation of the normalizing constant in the posterior distribution.

We distinguish two cases. First, suppose that the prior density, to be denoted here by $p(\theta)$, is continuous and nonzero in the neighbourhood of the true value θ^*, and, for simplicity, bounded elsewhere. The posterior density of Θ is then

$$p(\theta)\exp\{l(\theta)\} \Big/ \int p(\omega)\exp\{l(\omega)\}d\omega. \qquad (6.39)$$

Now expand $l(\omega)$ about $\hat{\theta}$ as in (6.21). Only values of θ and ω near $\hat{\theta}$ make an effective contribution, and the numerator is approximately

$$p(\hat{\theta})\exp\{l(\hat{\theta})\}\exp\{-(\theta-\hat{\theta})^2\hat{j}/2\}, \qquad (6.40)$$

whereas the denominator is, after some calculation, to the same order of approximation

$$(2\pi)^{1/2}p(\hat{\theta})\exp\{l(\hat{\theta})\}/\sqrt{\hat{j}}. \qquad (6.41)$$

That is, the posterior distribution is asymptotically normal with mean $\hat{\theta}$ and variance $\hat{j}^{-1}$. The approximation technique used here is called Laplace expansion. The asymptotic argument considered amounts to supposing that the prior varies little over the range of values of θ reasonably consistent with the data as judged by the likelihood function. As with all other asymptotic considerations, the extent to which the assumptions about the prior and the likelihood are justified needs examination in each application.

The caution implicit in the last statement is particularly needed in the second case that we now consider where there is an atom of prior probability π_0 at a null hypothesis $\theta = \theta_0$ and the remaining prior probability is distributed over the nonnull values. It is tempting to write this latter part in the form $(1-\pi_0)p_0(\theta)$, where $p_0(\theta)$ is some smooth density not depending on n. This is, however, often to invite misinterpretation, because in most, if not all, specific applications in which a test of such a hypothesis is thought worth doing, the only serious possibilities needing consideration are that either the null hypothesis is (very nearly) true or that some alternative within a range fairly close to θ_0 is true. This suggests that the remaining part of the prior density should usually be taken in the form $q\{(\theta-\theta_0)\sqrt{n}\}\sqrt{n}$, where $q(.)$ is some fixed probability density function. This is of scale and location form centred on θ_0 and with a dispersion of order $1/\sqrt{n}$. As $n \to \infty$ this concentrates its mass in a local region around θ_0. It then follows that the atom of posterior probability at θ_0 is such that the odds against θ_0, namely $\{1 - p(\theta_0 \mid y)\}/p(\theta_0 \mid y)$, are

$$\frac{(1 - \pi_0) \int \exp\{l(\theta)\} q\{(\theta - \theta_0)\sqrt{n}\}\sqrt{n} d\theta}{\pi_0 \exp\{l(\theta_0)\}}. \qquad (6.42)$$

Now expand the log likelihood around θ_0, replacing $j(\theta_0)$ by $n\bar{i}$ and the signed gradient statistic by $T_U = U(\theta_0; Y)/\sqrt{\{n\bar{i}(\theta_0)\}}$. The posterior odds against θ_0 are then approximately

$$(1 - p_0) \int \exp(vT_U - v^2/2) q(v/\sqrt{\bar{i}})/\sqrt{\bar{i}} dv/p_0. \qquad (6.43)$$

Thus as $n \to \infty$ the posterior odds are asymptotically a fixed function of the test statistic T_U, or indeed of one of the other essentially equivalent statistics studied above. That is, for a fixed $q(.)$ the relation between the significance level and the posterior odds is independent of n. This is by contrast with the theory with a fixed prior in which case the corresponding answer depends explicitly on n because, typically unrealistically, large portions of prior probability are in regions remote from the null hypothesis relative to the information in the data.

6.3 Multidimensional parameter

The extension of the results of the previous section when the parameter θ is d-dimensional is immediate. The gradient U is replaced by the $d \times 1$ gradient vector

$$U(\theta; Y) = \nabla l(\theta; Y), \qquad (6.44)$$

where $\nabla = (\partial/\partial \theta_1, \ldots, \partial/\partial \theta_d)^T$ forms a column vector of partial derivatives. Arguing component by component, we have as before that

$$E\{U(\theta; Y); \theta\} = 0. \qquad (6.45)$$

Then, differentiating the rth component of this equation with respect to θ_s of Θ, we have in generalization of (6.7) that the covariance matrix of $U(\theta; Y)$ is

$$\text{cov}\{U(\theta; Y); \theta\} = E\{U(\theta; Y)U^T(\theta; Y)\} = i(\theta), \qquad (6.46)$$

where the $d \times d$ matrix $i(\theta)$, called the *expected information matrix*, is the matrix of second derivatives

$$i(\theta) = E\{-\nabla \nabla^T l(\theta; Y); \theta\}. \qquad (6.47)$$

Because this is the covariance matrix of a random vector it is positive definite unless there are linear constraints among the components of the gradient vector and we assume these absent.

Further,

$$E\{U(\theta;Y);\theta+\delta\} = i(\theta)\delta + O(\|\delta\|^2). \qquad (6.48)$$

Suppose that we transform from θ by a $(1,1)$ transformation to new parameters $\phi(\theta)$, the transformation having Jacobian $\partial\phi/\partial\theta$ in which the rows correspond to the components of ϕ and the columns to those of θ. Adopting the notation used in the scalar parameter case, we have that the score vectors are related by

$$U^\Phi(\phi;Y) = (\partial\theta/\partial\phi)^T U^\Theta(\theta;Y) \qquad (6.49)$$

and the information matrices by

$$i^\Phi(\phi) = \frac{\partial\theta^T}{\partial\phi} i^\Theta(\theta) \frac{\partial\theta}{\partial\phi}. \qquad (6.50)$$

The inverse information matrices satisfy

$$\{i^\Phi(\phi)\}^{-1} = \frac{\partial\phi^T}{\partial\theta} \{i^\Theta(\theta)\}^{-1} \frac{\partial\phi}{\partial\theta}. \qquad (6.51)$$

These are the direct extensions of the results of Example 6.3.

In terms of asymptotics, we suppose in particular that $U(\theta;Y)$ has an asymptotic multivariate normal distribution and that $i(\theta)/n$ tends to a positive definite matrix $\bar{i}(\theta)$. The univariate expansions are replaced by their multivariate forms, for example expanding about the true value θ^*, so that we can write $l(\theta;Y)$ in the form

$$l(\theta^*;Y) + (\theta-\theta^*)^T U(\theta^*;Y) - (\theta-\theta^*)^T j(\theta^*)(\theta-\theta^*)/2, \qquad (6.52)$$

where $j(\theta^*)$ is minus the matrix of observed second derivatives.

Essentially the same argument as before now shows that $\hat{\theta}$ is asymptotically normal with mean θ^* and covariance matrix $i^{-1}(\theta^*)$ which can be replaced by $j^{-1}(\theta^*)$ or more commonly by $\hat{j}^{-1}$, the observed inverse information matrix at the maximum likelihood point.

Now if Z is normally distributed with zero mean and variance v, then $Zv^{-1}Z = Z^2/v$ has, by definition, a chi-squared distribution with one degree of freedom. More generally if Z is a $d \times 1$ vector having a multivariate normal distribution with zero mean and covariance matrix V of full rank d, then $\|Z\|_V^2 = Z^T V^{-1} Z$ has a chi-squared distribution with d degrees of freedom. To see this, note that the quadratic form is invariant under nonsingular transformations of Z

with corresponding changes in the covariance matrix and that Z can always be transformed to d independent standardized normal random variables. See also Note 2.6.

It follows that the pivots

$$(\theta - \hat{\theta})^T i^{-1}(\theta)(\theta - \hat{\theta}) \tag{6.53}$$

or

$$(\theta - \hat{\theta})^T \hat{j}^{-1}(\theta - \hat{\theta}) \tag{6.54}$$

can be used to form approximate confidence regions for θ. In particular, the second, and more convenient, form produces a series of concentric similar ellipsoidal regions corresponding to different confidence levels.

The three quadratic statistics discussed in Section 6.3.4 take the forms, respectively

$$W_E = (\hat{\theta} - \theta_0)^T i(\theta_0)(\hat{\theta} - \theta_0), \tag{6.55}$$

$$W_L = 2\{l(\hat{\theta}) - l(\theta_0)\}, \tag{6.56}$$

$$W_U = U(\theta_0; Y)^T i^{-1}(\theta_0) U(\theta_0; Y). \tag{6.57}$$

Again we defer discussion of the relative merits of these until Sections 6.6 and 6.11.

6.4 Nuisance parameters

6.4.1 The information matrix

In the great majority of situations with a multidimensional parameter θ, we need to write $\theta = (\psi, \lambda)$, where ψ is the parameter of interest and λ the nuisance parameter. Correspondingly we partition $U(\theta; Y)$ into two components $U_\psi(\theta : Y)$ and $U_\lambda(\theta; Y)$. Similarly we partition the information matrix and its inverse in the form

$$i(\theta) = \begin{pmatrix} i_{\psi\psi} & i_{\psi\lambda} \\ i_{\lambda\psi} & i_{\lambda\lambda} \end{pmatrix}, \tag{6.58}$$

and

$$i^{-1}(\theta) = \begin{pmatrix} i^{\psi\psi} & i^{\psi\lambda} \\ i^{\lambda\psi} & i^{\lambda\lambda} \end{pmatrix}. \tag{6.59}$$

There are corresponding partitions for the observed information $\hat{j}$.

6.4.2 Main distributional results

A direct approach for inference about ψ is based on the maximum likelihood estimate $\hat{\psi}$ which is asymptotically normal with mean ψ and covariance matrix $i^{\psi\psi}$ or equivalently $\hat{j}^{\psi\psi}$. In terms of a quadratic statistic, we have for testing whether $\psi = \psi_0$ the form

$$W_E = (\hat{\psi} - \psi_0)^T (i^{\psi\psi})^{-1} (\hat{\psi} - \psi_0) \tag{6.60}$$

with the possibility of using $(\hat{j}^{\psi\psi})^{-1}$ rather than $(i^{\psi\psi})^{-1}$. Note also that even if ψ_0 were used in the calculation of $i^{\psi\psi}$ it would still be necessary to estimate λ except for those special problems in which the information does not depend on λ.

Now to study ψ via the gradient vector, as far as possible separated from λ, it turns out to be helpful to write U_ψ as a linear combination of U_λ plus a term uncorrelated with U_λ, i.e., as a linear least squares regression plus an uncorrelated deviation. This representation is

$$U_\psi = i_{\psi\lambda} i_{\lambda\lambda}^{-1} U_\lambda + U_{\psi \cdot \lambda}, \tag{6.61}$$

say, where $U_{\psi \cdot \lambda}$ denotes the deviation of U_ψ from its linear regression on U_λ. Then a direct calculation shows that

$$\mathrm{cov}(U_{\psi \cdot \lambda}) = i_{\psi\psi \cdot \lambda}, \tag{6.62}$$

where

$$i_{\psi\psi \cdot \lambda} = i_{\psi\psi} - i_{\psi\lambda} i_{\lambda\lambda}^{-1} i_{\lambda\psi} = (i^{\psi\psi})^{-1}. \tag{6.63}$$

The second form follows from a general expression for the inverse of a partitioned matrix.

A further property of the adjusted gradient which follows by direct evaluation of the resulting matrix products by (6.63) is that

$$E(U_{\psi \cdot \lambda}; \theta + \delta) = i_{\psi\psi \cdot \lambda} \delta_\psi + O(\|\delta\|^2), \tag{6.64}$$

i.e., to the first order the adjusted gradient does not depend on λ. This has the important consequence that in using the gradient-based statistic to test a null hypothesis $\psi = \psi_0$, namely

$$W_U = U_{\psi \cdot \lambda}^T (\psi_0, \lambda) i^{\psi\psi}(\psi_0, \lambda) U_{\psi \cdot \lambda}(\psi_0, \lambda), \tag{6.65}$$

it is enough to replace λ by, for example, its maximum likelihood estimate given ψ_0, or even by inefficient estimates.

The second version of the quadratic statistic (6.56), corresponding more directly to the likelihood function, requires the collapsing of the log likelihood into a function of ψ alone, i.e., the elimination of dependence on λ.

This might be achieved by a semi-Bayesian argument in which λ but not ψ is assigned a prior distribution but, in the spirit of the present discussion, it is done by maximization. For given ψ we define $\hat{\lambda}_\psi$ to be the maximum likelihood estimate of λ and then define the profile log likelihood of ψ to be

$$l_P(\psi) = l(\psi, \hat{\lambda}_\psi), \tag{6.66}$$

a function of ψ alone and, of course, of the data. The analogue of the previous likelihood ratio statistic for testing $\psi = \psi_0$ is now

$$W_L = 2\{l_P(\hat{\psi}) - l_P(\psi_0)\}. \tag{6.67}$$

Expansions of the log likelihood about the point (ψ_0, λ) show that in the asymptotic expansion, we have to the first term that $W_L = W_U$ and therefore that W_L has a limiting chi-squared distribution with d_ψ degrees of freedom when $\psi = \psi_0$. Further because of the relation between significance tests and confidence regions, the set of values of ψ defined as

$$\{\psi : 2\{l_P(\hat{\psi}) - l_P(\psi)\} \leq k^{*2}_{d_\psi; c}\} \tag{6.68}$$

forms an approximate $1 - c$ level confidence set for ψ.

6.4.3 More on profile likelihood

The possibility of obtaining tests and confidence sets from the profile log likelihood $l_P(\psi)$ stems from the relation between the curvature of $l_P(\psi)$ at its maximum and the corresponding properties of the initial log likelihood $l(\psi, \lambda)$.

To see this relation, let ∇_ψ and ∇_λ denote the $d_\psi \times 1$ and $d_\lambda \times 1$ operations of partial differentiation with respect to ψ and λ respectively and let D_ψ denote total differentiation of any function of ψ and $\hat{\lambda}_\psi$ with respect to ψ. Then, by the definition of total differentiation,

$$D_\psi^T l_P(\psi) = \nabla_\psi^T l(\psi, \hat{\lambda}_\psi) + \nabla_\lambda^T l(\psi, \hat{\lambda}_\psi) D_\psi(\hat{\lambda}_\psi^T)^T. \tag{6.69}$$

Now apply D_ψ again to get the Hessian matrix of the profile likelihood in the form

$$D_\psi D_\psi^T l_P(\psi) = \nabla_\psi \nabla_\psi^T l(\psi, \hat{\lambda}_\psi) + \{\nabla_\psi \nabla_\lambda^T l(\psi, \hat{\lambda}_\psi)(\nabla_\psi \hat{\lambda}_\psi^T)\}^T$$
$$+ \{\nabla_\lambda^T l(\psi, \hat{\lambda}_\psi)\}\{\nabla_\psi \nabla_\psi^T \hat{\lambda}_\psi^T\} + (\nabla_\psi \hat{\lambda}^T)\nabla_\lambda \nabla_\psi^T l(\psi, \hat{\lambda}_\psi)$$
$$+ (\nabla_\psi \hat{\lambda}_\psi^T)\{\nabla_\lambda \nabla_\lambda^T l(\psi, \hat{\lambda}_\psi)(\nabla_\psi \hat{\lambda}_\psi^T)^T\}. \tag{6.70}$$

The maximum likelihood estimate $\hat{\lambda}_\psi$ satisfies for all ψ the equation $\nabla_\lambda^T l(\psi, \hat{\lambda}_\psi) = 0$. Differentiate totally with respect to ψ to give

$$\nabla_\psi \nabla_\lambda^T l(\psi, \hat{\lambda}_\psi) + (D_\psi \hat{\lambda}_\psi^T)\nabla_\lambda \nabla_\lambda^T l(\psi, \hat{\lambda}_\psi) = 0. \tag{6.71}$$

Thus three of the terms in (6.70) are equal except for sign and the third term is zero in the light of the definition of $\hat{\lambda}_\psi$. Thus, eliminating $D_\psi \hat{\lambda}_\psi$, we have that the formal observed information matrix calculated as minus the Hessian matrix of $l_P(\psi)$ evaluated at $\hat{\psi}$ is

$$\hat{j}_{P,\psi\psi} = \hat{j}_{\psi\psi} - \hat{j}_{\psi\lambda}\hat{j}_{\lambda\lambda}^{-1}\hat{j}_{\lambda\psi} = \hat{j}_{\psi\psi\cdot\lambda}, \qquad (6.72)$$

where the two expressions on the right-hand side of (6.72) are calculated from $l(\psi, \lambda)$. Thus the information matrix for ψ evaluated from the profile likelihood is the same as that evaluated via the full information matrix of all parameters.

This argument takes an especially simple form when both ψ and λ are scalar parameters.

6.4.4 Parameter orthogonality

An interesting special case arises when $i_{\lambda\psi} = 0$, so that approximately $j_{\lambda\psi} = 0$. The parameters are then said to be *orthogonal*. In particular, this implies that the corresponding maximum likelihood estimates are asymptotically independent and, by (6.71), that $D_\psi \hat{\lambda}_\psi = 0$ and, by symmetry, that $D_\lambda \hat{\psi}_\lambda = 0$. In nonorthogonal cases if ψ changes by $O(1/\sqrt{n})$, then $\hat{\lambda}_\psi$ changes by $O_p(1/\sqrt{n})$; for orthogonal parameters, however, the change is $O_p(1/n)$. This property may be compared with that of orthogonality of factors in a balanced experimental design. There the point estimates of the main effect of one factor, being contrasts of marginal means, are not changed by assuming, say, that the main effects of the other factor are null. That is, the $O_p(1/n)$ term in the above discussion is in fact zero.

There are a number of advantages to having orthogonal or nearly orthogonal parameters, especially component parameters of interest. Independent errors of estimation may ease interpretation, stability of estimates of one parameter under changing assumptions about another can give added security to conclusions and convergence of numerical algorithms may be speeded. Nevertheless, so far as parameters of interest are concerned, subject-matter interpretability has primacy.

Example 6.4. *Mixed parameterization of the exponential family.* Consider a full exponential family problem in which the canonical parameter ϕ and the canonical statistic s are partitioned as (ϕ_1, ϕ_2) and (s_1, s_2) respectively, thought of as column vectors. Suppose that ϕ_2 is replaced by η_2, the corresponding component of the mean parameter $\eta = \nabla k(\phi)$, where $k(\phi)$ is the cumulant generating function occurring in the standard form for the family. Then ϕ_1 and η_2 are orthogonal.

6.4 Nuisance parameters

To prove this, we find the Jacobian matrix of the transformation from (ϕ_1, η_2) to (ϕ_1, ϕ_2) in the form

$$\partial(\phi_1, \eta_2)/\partial(\phi_1, \phi_2) = \begin{pmatrix} I & 0 \\ \nabla_1 \nabla_2^T k(\phi) & \nabla_2 \nabla_2^T k(\phi) \end{pmatrix}. \tag{6.73}$$

Here ∇_l denotes partial differentiation with respect to ϕ_l for $l = 1, 2$.

Combination with (6.51) proves the required result. Thus, in the analysis of the 2×2 contingency table the difference of column means is orthogonal to the log odds ratio; in a normal distribution mean and variance are orthogonal.

Example 6.5. *Proportional hazards Weibull model.* For the Weibull distribution with density

$$\gamma \rho (\rho y)^{\gamma - 1} \exp\{-(\rho y)^\gamma\} \tag{6.74}$$

and survivor function, or one minus the cumulative distribution function,

$$\exp\{-(\rho y)^\gamma\}, \tag{6.75}$$

the hazard function, being the ratio of the two, is $\gamma \rho (\rho y)^{\gamma - 1}$.

Suppose that $Y_1, \ldots, Y_n$ are independent random variables with Weibull distributions all with the same γ. Suppose that there are explanatory variables $z_1, \ldots, z_n$ such that the hazard is proportional to $e^{\beta z}$. This is achieved by writing the value of the parameter ρ corresponding to Y_k in the form $\exp\{(\alpha + \beta z_k)/\gamma\}$. Here without loss of generality we take $\Sigma z_k = 0$. In many applications z and β would be vectors but here, for simplicity, we take one-dimensional explanatory variables. The log likelihood is

$$\Sigma(\log \gamma + \alpha + \beta z_k) + (\gamma - 1)\Sigma \log y_k - \Sigma \exp(\alpha + \beta z_k) y_k^\gamma. \tag{6.76}$$

Direct evaluation now shows that, in particular,

$$E\left(-\frac{\partial^2 l}{\partial \alpha \partial \beta}\right) = 0, \tag{6.77}$$

$$E\left(-\frac{\partial^2 l}{\partial \beta \partial \gamma}\right) = -\frac{\beta \Sigma z_k^2}{\gamma}, \tag{6.78}$$

$$E\left(-\frac{\partial^2 l}{\partial \gamma \partial \alpha}\right) = -\frac{0.5771 + \alpha}{\gamma}, \tag{6.79}$$

where Euler's constant, 0.5771, arises from the integral

$$\int_0^\infty v \log v \, dv. \tag{6.80}$$

Now locally near $\beta = 0$ the information elements involving β are zero or small implying local orthogonality of β to the other parameters and in particular to γ. Thus not only are the errors of estimating β almost uncorrelated with those of the other parameters but, more importantly in some respects, the value of $\hat{\beta}_\gamma$ will change only slowly with γ. In some applications this may mean that analysis based on the exponential distribution, $\gamma = 1$, is relatively insensitive to that assumption, at least so far as the value of the maximum likelihood estimate of β is concerned.

6.5 Tests and model reduction

One of the uses of these results for vector parameters of interest is in somewhat exploratory investigation of the form of model to be used as a basis for interpretation steered by a wish, for various reasons, to use reasonably simple models wherever feasible. We may have in addition to common nuisance parameters that the parameter of interest ψ can be partitioned, say as $(\psi_1^T, \ldots, \psi_h^T)$, and that models with only the first g components, $g < h$, may be preferred if reasonably consistent with the data. The individual components will typically be vectors; it is also not essential to assume, as here, that the model reductions considered are hierarchical.

We can thus produce a series of maximized log likelihoods involving maximization always over λ taken with the full vector ψ, the vector ψ with $\psi_h = 0$, down to the vector ψ_1 alone, or possibly even with $\psi = 0$. If we denote these maximized log likelihoods by $\hat{l}_h, \ldots, \hat{l}_0$ a formal test that, say, $\psi_g = 0$ given that $\psi_{g+1} = \cdots = \psi_h = 0$ is provided by

$$2(\hat{l}_g - \hat{l}_{g-1}) \tag{6.81}$$

tested by comparison with the chi-squared distribution with degrees of freedom the dimension of ψ_g.

This generates a sequence of values of chi-squared and in a sense the hope is that this sequence jumps suddenly from the highly significant to small values, all consistent with the relevant degrees of freedom, indicating unambiguously the simplest model in the sequence reasonably consistent with the data.

If the models being considered are not nested within some overall model the analysis is more complicated. Suppose that we have two distinct families of models $f(y; \theta)$ and $g(y; \phi)$ which can give sufficiently similar results that choice between them may be delicate, but which are *separate* in the sense that any particular distribution in one family cannot be recovered as a special case of the other. A simple example is where there are n independent and identically

6.5 Tests and model reduction

distributed components, with one family being exponential and the other log normal, so that θ is one-dimensional and ϕ is two-dimensional.

There are broadly three approaches to such problems. One is to produce a comprehensive model reducing to the two special cases when a specifying parameter takes values, say 0 and 1. Such a family might be

$$c(\psi, \theta, \phi) \exp\{\psi \log f(y; \theta) + (1 - \psi) \log g(y; \phi)\}, \quad (6.82)$$

used to estimate ψ and very particularly to examine possible consistency with the defining values 0 and 1.

A simpler and in many ways the most satisfactory approach is to follow the prescription relating confidence intervals to significance tests. That is, we take each model in turn as the null hypothesis and test, for example by the log likelihood ratio $\log\{f(y; \hat\theta)/g(y; \hat\phi)\}$, consistency first with f and then with g. This leads to the conclusion that both, either one, or neither model is reasonably consistent with the data at whatever significance levels arise. The distributional results obtained above for the nested case no longer apply; in particular the log likelihood ratio test statistic has support on positive and negative real values. While analytical results for the distribution are available, much the most effective procedure is simulation.

A third approach is Bayesian and for this we suppose that the prior probabilities of the two models are respectively $\pi_f, \pi_g = 1 - \pi_f$ and that the conditional prior densities of the parameters given the model are respectively $p_f(\theta)$ and $p_g(\phi)$. Then the ratio of the posterior probability of model g to that of model f, i.e., the odds ratio of g versus f, is

$$\frac{\pi_g \int \exp\{l_g(\phi)\} p_g(\phi) d\phi}{\pi_f \int \exp\{l_f(\theta)\} p_f(\theta) d\theta}, \quad (6.83)$$

where l_g and l_f are the relevant log likelihoods. Separate Laplace expansion of numerator and denominator gives that the odds ratio is approximately

$$\frac{\pi_g \exp(\hat l_g) p_g(\hat\phi)(2\pi)^{d_g/2} \hat\Delta_g^{-1/2}}{\pi_f \exp(\hat l_f) p_f(\hat\theta)(2\pi)^{d_f/2} \hat\Delta_f^{-1/2}}. \quad (6.84)$$

Here, for example, d_g is the dimension of ϕ and $\hat\Delta_g$ is the determinant of the observed information matrix in model g.

A crucial aspect of this is the role of the prior densities of the parameters within each model. In a simpler problem with just one model a flat or nearly flat prior density for the parameter will, as we have seen, disappear from the answer, essentially because its role in the normalizing constant for the posterior density cancels with its role in the numerator. This does not happen here. Numerator and denominator both have terms expressing the ratio of the maximized likelihood to that of the approximating normal density of the maximum likelihood estimate. Entry of the prior densities is to be expected. If one model has parameter estimates in a region of relatively high prior density that is evidence in favour of that model.

To proceed with a general argument, further approximations are needed. Suppose first that the parameter spaces of the two models are of equal dimension, $d_g = d_f$. We can then set up a (1, 1) correspondence between the two parameter spaces, for example by making a value ϕ' correspond to the probability limit of $\hat{\theta}$ when calculated from data distributed according to $g(y; \phi')$. Then, if the prior density $p_f(\theta)$ is determined from $p_g(\phi)$ by the same transformation and the determinants of the observed information related similarly, it follows that the posterior odds are simply

$$\frac{\pi_g \exp(\hat{l}_g)}{\pi_f \exp(\hat{l}_f)} \qquad (6.85)$$

and to this level of approximation the discussion of Example 5.13 of the comparison of completely specified models applies with maximized likelihood replacing likelihood. The assumptions about the prior densities are by no means necessarily true; it is possible that there is substantial prior information about ϕ if g is true and only weak prior information about θ if f is true.

When, say, $d_g > d_f$ the argument is even more tentative. Suppose that ϕ can be partitioned into two parts, one of dimension d_f that can be matched to θ and the other of dimension $d_g - d_f$, which is in some sense approximately orthogonal to the first part and whose probability limit does not depend on θ. The contribution of the latter part can come only from the prior and we suppose that it is equivalent to that supplied by n_0 fictitious observations, as compared with the n real observations we have for analysis. Then the ratio of the prior density to the corresponding information contribution is approximately $(n_0/n)^{(d_g - d_f)/2}$.

This indicates the penalty to be attached to a maximized likelihood for every additional component fitted. This is most concisely represented by replacing a maximized log likelihood, say $\hat{l}_g$ by

$$\hat{l}_g - (d_g/2) \log n. \qquad (6.86)$$

The term in n_0 has been omitted as being small compared with n, although the logarithmic dependence on n is warning that in any case the approximation may be poor.

This is called the *Bayesian information criterion*, BIC, although it was originally suggested from a frequentist argument. A number of somewhat similar criteria have been developed that are based on the $\hat{l}_g$, but all of which aim to provide an automatic criterion, not based on significance-testing considerations, for the selection of a model. They all involve penalizing the log likelihood for the number of parameters fitted. The relevance of any relatively automatic method of model selection depends strongly on the objective of the analysis, for example as to whether it is for explanation or for empirical prediction. To discuss the reasonableness of these and similar procedures would involve detailed examination of the strategy to be used in applying statistical methods. This is defined to be outside the scope of the present book.

For interpretative purposes no automatic method of model selection is appropriate and while the sequence of tests sketched above may often give some useful guidance, mechanical application is likely to be unwise.

If the Bayesian formulation is adopted and if the posterior distribution of all aspects is reasonably defined a procedure of *Bayesian model averaging* is formally possible. If, for example, the objective is predicting the value of y corresponding to some specific values of explanatory variables, then in essence predictions from the separate models are weighted by the relevant posterior probabilities. For the method to be appropriate for interpretation, it is, however, essential that the parameter of interest ψ is defined so as to have identical interpretations in the different models and this will often not be possible.

6.6 Comparative discussion

There are thus three broad routes to finding asymptotic test and confidence regions, two with variants depending on whether observed or expected information is used and at what point the information is evaluated. In the one-dimensional case there is also the issue of whether equi-tailed intervals are appropriate and this has already been discussed in Section 3.3. In most cases the numerical differences between the three key asymptotic procedures are of no consequence, but it is essential to discuss what to do when there are appreciable discrepancies.

Were the observed log likelihood exactly quadratic around θ the procedures using $\hat{j}$ as the basis for assessing precision would be exactly the same for the three procedures. Procedures based directly on the maximum likelihood estimate and a measure of precision have substantial advantages for the concise summarization of conclusions via an estimate and its estimated precision which, via the implied pivot, can be used both for tests and confidence intervals and as a base for further analysis, combination with other data and so on.

They do, however, have the disadvantage, substantial under some circumstances, of not being invariant under nonlinear transformations of θ; the other two statistics do have this exact invariance. Indeed W_L from (6.56) can often be regarded as close to the value that W_E from (6.55) with $\hat{j}$ would take after reparameterization to near-quadratic form. There can be additional and related difficulties with W_E if $j(\theta)$ or $i(\theta)$ changes rapidly with θ in the relevant range.

Two further issues concern the adequacy of the normal or chi-squared form used to approximate the distribution of the test statistics and behaviour in very nonstandard situations. We return to these matters briefly in Section 6.11 and Chapter 7. Both considerations tend to favour the use of W_L, the likelihood ratio statistic, and a simple summary of the arguments is that this is the safest to use in particular applications should any discrepancies appear among the different statistics, although in extremely nonstandard cases special analysis is, in principle, needed.

For methods involving the maximum likelihood estimate and an estimate of its standard error the preference for staying close to a direct use of the likelihood, as well as more elaborate arguments of the kind summarized in the next section, suggest a preference for using the observed information $\hat{j}$ rather than the expected information $i(\hat{\theta})$. The following examples reinforce that preference.

Example 6.6. *A right-censored normal distribution.* Let $Y_1, \ldots, Y_n$ be independently normally distributed with unknown mean μ and variance σ_0^2, assumed for simplicity to be known. Suppose, however, that all observations greater than a known constant b are recorded simply as being greater than b, their individual values not being available. The log likelihood is

$$-\Sigma'(y_k - \mu)^2/(2\sigma_0^2) + r \log \Phi\{(\mu - a)/\sigma_0\}, \qquad (6.87)$$

where r is the number of censored observations and Σ' denotes summation over the uncensored observations.

The sufficient statistic is thus formed from the mean of the uncensored observations and the number, r, censored. There is no exact ancillary statistic suitable for conditioning. The maximum likelihood estimating equation is

$$\hat{\mu} = \bar{y}' + r\sigma_0(n-r)^{-1}v\{(\mu - a)/\sigma_0\}, \qquad (6.88)$$

where $v(x) = \Phi'(x)/\Phi(x)$ is the reciprocal of the Mills ratio. Moreover, the observed information is given by

$$\hat{j}\sigma_0^2 = n - r - rv'\{(\hat{\mu} - a)/\sigma_0\}. \qquad (6.89)$$

An important point is that if $r = 0$, the inference is unaffected, to the first order at least, by the possibility that censoring might have occurred but in fact did not; the standard pivot for the mean of a normal distribution is used.

This is in line with the Bayesian solution that, except possibly for any impact of potential censoring on the prior, the inference for $r = 0$ would also be unaffected. Were the expected information to be used instead of the observed information, this simple conclusion would not hold, although in most cases the numerical discrepancy between the two pivots would be small.

The following is a more striking example of the need to consider observed rather than expected information.

Example 6.7. *Random walk with an absorbing barrier.* Consider a random walk in which if at time m the walk is at position k, the position at time $m + 1$ is $k + 1$ with probability θ and is $k - 1$ with probability $1 - \theta$, the steps at different times being mutually independent. Suppose that at time zero the walk starts at point y_0, where $y_0 > 0$. Suppose, further, that there is an absorbing barrier at zero, so that if the walk ever reaches that point it stays there for ever.

The following qualitative properties of the walk are relevant. If $\theta \leq 1/2$ absorption at zero is certain. If $\theta > 1/2$ there are two distinct types of path. With probability $\{(1 - \theta)/\theta\}^{y_0}$ the walk ends with absorption at zero, whereas with the complementary probability the walk continues indefinitely reaching large values with high probability.

If one realization of the walk is observed for a long time period, the likelihood is $\theta^{r_+}(1 - \theta)^{r_-}$, where r_+ and r_- are respectively the numbers of positive and negative steps. Use of the expected information would involve, when $\theta > 1/2$, averaging over paths to absorption and paths that escape to large values and would be misleading, especially if the data show speedy absorption with little base for inferring the value of θ. Note though that the event of absorption does not provide an ancillary statistic.

A formal asymptotic argument will not be given but the most interesting case is probably to allow observation to continue for a large number n of steps, and to suppose that $y_0 = b_0\sqrt{n}$ and that $\theta = 1/2 + \delta/\sqrt{n}$.

A very similar argument applies to the Markovian birth–death process in continuous time, generalizing the binary fission process of Example 2.4. In this particles may die as well as divide and if the number of particles reaches zero the system is extinct.

6.7 Profile likelihood as an information summarizer

We have seen in the previous section some preference for profile log likelihood as a basis for inference about the parameter of interest ψ over asymptotically equivalent bases, in particular over the maximum likelihood estimate $\hat{\psi}$ and its estimated covariance matrix $\hat{j}^{\psi\psi}$. If the profile log likelihood is essentially symmetrical locally around $\hat{\psi}$ the approaches are nearly equivalent and, especially if ψ is one-dimensional, specification of the maximum likelihood estimate and its standard error is more directly interpretable and hence preferable. While a parameter transformation may often be used, in effect to correct for asymmetry in the profile log likelihood, as already stressed physical interpretability of a parameter must take preference over the statistical properties of the resulting estimates and so parameter transformation may not always be appropriate.

An important role for statistical analysis is not only the clarification of what can reasonably be learned from the data under analysis, but also the summarization of conclusions in a form in which later work can use the outcome of the analysis, for example in combination with subsequent related results. Recourse to the original data may sometimes be impracticable and, especially in complex problems, only information about the parameter ψ of interest may be worth retaining.

Where the profile log likelihood $l_P(\psi)$ is appreciably asymmetric around the maximum, or in other cases of notable departure from local quadratic form, it may be sensible to record either the profile log likelihood in numerical form or to record say the third and even fourth derivatives at the maximum, as well as the second, the information element. When independent sets of data potentially relevant to ψ become available the overall profile likelihood can then be reconstructed as the sum of the separate contributions. The mutual consistency of the various estimates of ψ can be examined via a chi-squared test and, subject to reasonable consistency, a pooled estimate, the overall maximum likelihood estimate, found. This analysis implicitly assumes that nuisance parameters in different sets of data are distinct or, if that is not the case, that the information lost by ignoring common elements is unimportant.

6.8 Constrained estimation

In general, the more parameters, the more complicated the estimation problem. There are exceptions, however, notably when a model is expanded to a saturated form or, more generally, where the (k, d) curved exponential family model is expanded to the full (k, k) family. We may sometimes usefully estimate within the larger family and then introduce the specific constraints that apply to the specific problem. This idea can, in particular, be used to find simple estimates asymptotically equivalent to maximum likelihood estimates. We shall relate the method more specifically to the exponential family.

6.8 Constrained estimation

Suppose that the model of interest is defined by a parameter vector θ and that the extended or covering model for which estimation is simpler is defined by $\phi = (\theta, \gamma)$. If $i(\phi)$, partitioned in the usual way, denotes the expected information matrix and $U_\phi^T = (U_\theta^T, U_\gamma^T)$ is the gradient of the log likelihood, the maximum likelihood estimates, $\hat{\theta}_r$, under the reduced, i.e., originating, model and the estimates, $\hat{\theta}_c$, under the covering model satisfy after local linearization

$$i_{\theta\theta}(\hat{\theta}_r - \theta) = U_\theta, \tag{6.90}$$

$$i_{\theta\theta}(\hat{\theta}_c - \theta) + i_{\theta\gamma}(\hat{\gamma}_c - \gamma) = U_\theta. \tag{6.91}$$

Suppose that γ is defined so that the model of interest corresponds to $\gamma = 0$. Then

$$\tilde{\theta}_r = \hat{\theta}_c + i_{\theta\theta}^{-1} i_{\theta\gamma} \hat{\gamma}_c \tag{6.92}$$

is close to $\hat{\theta}_r$, the difference being $O_p(1/n)$. The expected information may be replaced by the observed information. The estimate $\tilde{\theta}_r$ is essentially the estimate under the covering model adjusted for regression on $\hat{\gamma}_c$, which is an estimate of zero. If the regression adjustment were appreciable this would be an indication that the reduced model is inconsistent with the data.

Realistic applications of this idea are to relatively complex contingency table problems and to estimation of normal-theory covariance matrices subject to independence constraints. In the latter case the covering model specifies an unconstrained covariance matrix for which estimation is simple. The following is an artificial example which shows how the method works.

Example 6.8. *Curved exponential family model.* After reduction by sufficiency, suppose that $(\bar{Y}_1, \bar{Y}_2)$ are independently normally distributed with mean $(\theta, b\theta^2)$ and variances σ_0^2/n, where σ_0^2 is known. Here b is a known constant. This forms a (2, 1) exponential family and there is no ancillary statistic to reduce the problem further. Consider the covering model in which the mean is $(\theta, b\theta^2 + \gamma)$. Then the estimates under the covering model are $\hat{\theta}_c = \bar{Y}_1, \hat{\gamma}_c = \bar{Y}_2 - b\bar{Y}_1^2$. The information matrix at $\gamma = 0$ has elements

$$i_{\theta\theta} = (1 + 4b^2\theta^2)n/\sigma_0^2, \quad i_{\theta\gamma} = 2b\theta n/\sigma_0^2, \quad i_{\gamma\gamma} = n/\sigma_0^2. \tag{6.93}$$

Thus, on replacing θ in the matrix elements by the estimate $\bar{Y}_1$, we have that

$$\tilde{\theta}_r = \bar{Y}_1 + \frac{2b\bar{Y}_1}{(1 + 4b^2\bar{Y}_1^2)}(-b\bar{Y}_1^2 + \bar{Y}_2). \tag{6.94}$$

That is, the adjusted estimate differs from the simple estimate $\bar{Y}_1$ by an amount depending on how far the observations are from the curve defining the exponential family.

Direct calculation with the information matrices shows that, in contrast with $\mathrm{var}(\bar{Y}_1) = \sigma_0^2/n$, the new estimate has asymptotic variance

$$(\sigma_0^2/n)(1 + 4b^2\theta^2)^{-1}. \tag{6.95}$$

We now take a different route to closely related results. Consider the exponential family model with log likelihood

$$s^T\phi - k(\phi). \tag{6.96}$$

Recall that the mean parameter is $\eta = \nabla k(\phi)$, and that the information matrix $i(\theta)$ is the Jacobian of the transformation from ϕ to η.

Suppose that the model is constrained by requiring some components of the canonical parameter to be constant, often zero, i.e.,

$$\phi_c = 0 \quad (c \subset C). \tag{6.97}$$

We therefore consider a Lagrangian with Lagrange multipliers ν_c, namely

$$s^T\phi - k(\phi) - \Sigma_{c \subset C}\nu_c\phi_c. \tag{6.98}$$

If we differentiate with respect to ϕ_m for m not in C we have $s_m = \hat{\eta}_m$, whereas for $m \subset C$, we have that

$$s_m - \hat{\eta}_m - \nu_m = 0. \tag{6.99}$$

The Lagrange multipliers ν_m are determined by solving the simultaneous equations obtained by expressing ϕ as a function of η as

$$\phi_c(s_U, s_C - \nu) = 0, \tag{6.100}$$

where s_U are the components of the canonical statistic not in the conditioning set C. That is, the estimates of the mean parameters not involved in the constraints are unaffected by the constraints, whereas the estimates of the other mean parameters are defined by enforcing the constraints.

It can be shown that if standard n asymptotics holds, then ν is $O(1/\sqrt{n})$ and expansion of (6.100) yields essentially the noniterative procedure given at the beginning of this section.

A similar analysis, different in detail, applies if the constraints pertain to the mean parameter η rather than to the canonical parameter ϕ. In this case the Lagrangian is

$$s^T\phi - k(\phi) - \omega^T\eta_C, \tag{6.101}$$

with Lagrange multipliers ω.

Differentiate with respect to ϕ_m to give maximum likelihood estimating equations

$$s_m - \hat{\eta}_m - \Sigma \omega_k \partial \hat{\eta}_k / \partial \phi_m = 0, \qquad (6.102)$$

i.e.,

$$s_m - \hat{\eta}_m - \hat{\imath}_{km} \omega_k = 0. \qquad (6.103)$$

For $m \subset C$ we require $\hat{\eta}_m = 0$, so that

$$s_C = \hat{\imath}_{CC} \hat{\lambda}, \qquad (6.104)$$

where $\hat{\imath}_{CC}$ is the part of the information matrix referring to the variables in C and evaluated at the maximum likelihood point. Similarly, for the variables U not in C,

$$\hat{\eta}_U = s_U - \hat{\imath}_{PC} \hat{\imath}_{CC} s_C. \qquad (6.105)$$

If we have n replicate observations and standard n asymptotics apply we may again replace $\hat{\imath}$ by simpler estimates with errors $O_p(1/\sqrt{n})$ to obtain noniterative estimates of η_U that are asymptotically equivalent to maximum likelihood estimates.

Example 6.9. *Covariance selection model.* A special case, arising in particular in the discussion of Markovian undirected graphical systems, is based on independent and identically distributed $p \times 1$ random vectors normally distributed with covariance matrix Σ. There is no essential loss of generality in supposing that the mean is zero. The log likelihood for a single observation is

$$(\log \det \Sigma^{-1} - y^T \Sigma^{-1} y)/2 \qquad (6.106)$$

and the second term can be written

$$-\text{tr}\{\Sigma^{-1}(yy^T)\}/2. \qquad (6.107)$$

Here the trace, tr, of a square matrix is the sum of its diagonal elements.

It follows that for the full data the log likelihood is

$$\{n \log \det \Sigma^{-1} - \text{tr}(\Sigma^{-1} S)\}/2, \qquad (6.108)$$

where S is the matrix of observed sums of squares and products. The concentration matrix is the canonical parameter, the canonical statistic is S and the mean parameter $n\Sigma$. It follows that in the unconstrained model the maximum likelihood estimate of Σ, obtained by equating the canonical statistic to its expectation, is the sample covariance matrix T/n.

Now consider the constrained model in which the elements of Σ^{-1} in the set C are zero. The statistical interpretation is that the pairs of components in C are conditionally independent given the remaining components. It follows from the general discussion that the maximum likelihood estimates of the covariances of pairs of variables not in C are unchanged, i.e., are the sample covariances. The remaining estimates are defined by forcing the implied concentration elements to zero as required by the constraints. Approximations can be obtained by replacing the information matrices by large-sample approximations.

6.9 Semi-asymptotic arguments

In some applications relatively simple exact inference would be available about the parameter of interest ψ were the nuisance parameter λ known. If errors of estimating λ are relatively small the following argument may be used.

Suppose that to test the hypothesis $\psi = \psi_0$ for known λ the p-value corresponding to data y is a function of part of the data y_ψ and is $p(y_\psi, \psi_0; \lambda)$. Suppose that conditionally on y_ψ the estimate $\tilde{\lambda}$ of λ has expectation $\lambda + o(1/n)$ and covariance matrix $v(\lambda)/n$. Now taking expectations over the distribution of $\tilde{\lambda}$ given y_ψ, we have, after Taylor expansion to two terms, that

$$E\{p(y_\psi, \psi_0, \tilde{\lambda})\} = p(y_\psi, \psi_0, \lambda) + \frac{1}{2n}\text{tr}\{v(\lambda)\nabla_\lambda \nabla_\lambda^T p(y_\psi, \psi_0, \lambda)\}. \quad (6.109)$$

Now an estimated probability with expectation equal to a true value will itself, subject to lying in $(0, 1)$, have a probability interpretation. This leads to consideration of

$$p(y_\psi, \psi_0, \tilde{\lambda}) - (2n)^{-1}\text{tr}\{v(\tilde{\lambda})\nabla_\lambda \nabla_\lambda^T p(y_\psi, \psi_0, \tilde{\lambda})\}. \quad (6.110)$$

This is most safely expressed in the asymptotically equivalent form $p(y_\psi, \psi_0, \tilde{\lambda} + \tilde{\chi})$, where $\tilde{\chi}$ is any solution of

$$\tilde{\chi}\nabla_\lambda p = -(2n)^{-1}\text{tr}(v\nabla_\lambda \nabla_\lambda^T p). \quad (6.111)$$

Thus if λ is a scalar we take

$$p\left\{y_\psi, \psi_0, \tilde{\lambda} - \frac{v(\tilde{\lambda})\partial^2 p/\partial \lambda^2}{2n\partial p/\partial \lambda}\right\}. \quad (6.112)$$

Example 6.10. *Poisson-distributed signal with estimated background.* Suppose that in generalization of Example 3.7 we observe two random variables Y and Y_B having independent Poisson distributions with means respectively $(\psi + \lambda)t_S$ and λt_B. Thus Y refers to a signal plus background observed for a time t_S and Y_B to the background observed for a time t_B. We consider a situation in which t_B is long compared with t_S, so that the background is relatively precisely estimated. If λ were known the p-value for testing $\psi = \psi_0$ sensitive against larger values of ψ would be

$$p(y, \psi_0, \lambda) = \Sigma_{v=0}^{y} \exp\{-(\psi_0 + \lambda)t_S\}\{(\psi_0 + \lambda)t_S\}^v / v!. \qquad (6.113)$$

We write $\tilde{\lambda} = t_B^{-1} y_B$ and to a sufficient approximation $v(\lambda)/n = y_B/t_B^2$. In the special case when $y = 0$ the adjusted p-value becomes

$$\exp[-\psi_0 t_S - y_B t_S t_B^{-1} \{1 + t_S/(2t_B)\}]. \qquad (6.114)$$

6.10 Numerical-analytic aspects

6.10.1 General remarks

In the majority of applications the maximum of the likelihood $l(\theta)$ and associated statistics have to be found numerically or in the case of Bayesian calculations broadly comparable numerical integrations have to be done. In both frequentist and Bayesian approaches it may be fruitful to explore the general character of the likelihood function and, in particular, to compute and plot the profile log likelihood in the region of the maximum. The details of these procedures raise numerical-analytic rather than statistical issues and will not be discussed in depth here. We deal first with deterministic numerical methods and then with simulation-based approaches.

6.10.2 Numerical maximization

One group of techniques for numerical maximization of a function such as a log likelihood can be classified according as specification in readily computable form is possible for

- only function values,
- function values and first derivatives,
- function values and first and second derivatives.

In some situations an initial value of θ is available that is virtually sure to be close to the required maximum. In others it will be wise to start iterative algorithms from several or indeed many different starting points and perhaps also to use first a very coarse grid search in order to assess the general shape of the surface.

The most direct method available for low-dimensional problems is grid search, i.e., computation of $l(\theta)$ over a suitably chosen grid of points. It can be applied more broadly to the profile likelihood when ψ is, say, one- or two-dimensional, combined with some other method for computing $\hat{\lambda}_\psi$. Grid search, combined with a plot of the resulting function allows checking for likelihoods of nonstandard shapes, for example those departing appreciably from the quadratic form that is the basis of asymptotic theory.

In particular, a simple but effective way of computing the profile likelihood for one- or even two-dimensional parameters of interest, even in relatively complicated problems, is just to evaluate $l(\psi, \lambda)$ for a comprehensive grid of values of the argument; this in some contexts is feasible provided a reasonably efficient algorithm is available for computing l. The envelope of a plot of l against ψ shows the profile log likelihood. This is especially useful if ψ is a relatively complicated function of the parameters in terms of which the likelihood is most conveniently specified.

Next there is a considerable variety of methods of function maximization studied in the numerical-analytic literature, some essentially exploiting local quadratic approximations applied iteratively. They are distinguished by whether none, one or both of the first and second derivatives are specified analytically. All such procedures require a convergence criterion to assess when a maximum has in effect been reached. Especially when there is the possibility of a somewhat anomalous log likelihood surface, caution is needed in judging that convergence is achieved; reaching the same point from very different starting points is probably the best reassurance.

The most widely used method of the first type, i.e, using only function evaluations, is the Nelder–Mead simplex algorithm. In this the log likelihood is evaluated at the vertices of a simplex and further points added to the simplex in the light of the outcome, for example reflecting the point with lowest log likelihood in the complementary plane. There are numerous possibilities for changing the relative and absolute sizes of the edges of the simplex. When the algorithm is judged to have converged subsidiary calculations will be needed to assess the precision of the resulting estimates.

Many of the procedures are some variant of the Newton–Raphson iteration in which the maximum likelihood estimating equation is linearized around θ_0 to give

$$\nabla l(\theta_0) + \nabla \nabla^T l(\theta_0)(\hat{\theta} - \theta_0) = 0, \quad (6.115)$$

leading to the iterative scheme

$$\hat{\theta}_{t+1} - \hat{\theta}_t = -\{\nabla \nabla^T l(\hat{\theta}_t)\}^{-1} \nabla l(\hat{\theta}_t). \quad (6.116)$$

This requires evaluation of first and second derivatives. In many problems of even modest complexity the direct calculation of second derivatives can be time-consuming or impracticable so that indirect evaluation via a set of first derivatives is to be preferred. Such methods are called quasi-Newton methods.

These and related methods work well if the surface is reasonably close to quadratic in a region including both the starting point and the eventual maximum. In general, however, there is no guarantee of convergence.

An important technique for finding maximum likelihood estimates either when there is appreciable missing information which distorts a simple estimation scheme or where the likelihood is specified indirectly via latent variables involves what is sometimes called a self-consistency or missing information principle and is encapsulated in what is called the EM algorithm. The general idea is to iterate between formally estimating the missing information and estimating the parameters.

It can be shown that each step of the standard version of the EM algorithm can never decrease the log likelihood of Y. Usually, but not necessarily, this implies convergence of the algorithm, although this may be to a subsidiary rather than a global maximum and, in contrived cases, to a saddle-point of the log likelihood.

There are numerous elaborations, for example to speed convergence which may be slow, or to compute estimates of precision; the latter are not obtainable directly from the original EM algorithm.

In all iterative methods specification of when to stop is important. A long succession of individually small but steady increases in log likelihood is a danger sign!

6.10.3 Numerical integration

The corresponding numerical calculations in Bayesian discussions are ones of numerical integration arising either in the computation of the normalizing constant after multiplying a likelihood function by a prior density or by marginalizing a posterior density to isolate the parameter of interest.

The main direct methods of integration are Laplace expansion, usually taking further terms in the standard development of the expansion, and adaptive quadrature. In the latter standard procedures of numerical integration are used that are highly efficient when the function evaluations are at appropriate points. The adaptive part of the procedure is concerned with steering the choice of points sensibly, especially when many evaluations are needed at different parameter configurations.

6.10.4 Simulation-based techniques

The above methods are all deterministic in that once a starting point and a convergence criterion have been set the answer is determined, except for what are normally unimportant rounding errors. Such methods can be contrasted with simulation-based methods. One of the simplest of these in Bayesian discussions is importance sampling in which a numerical integration is interpreted as an expectation, which in turn is estimated by repeated sampling. The most developed type of simulation procedure is the Markov chain Monte Carlo, MCMC, method. In most versions of this the target is the posterior distribution of a component parameter. This is manoeuvred to be the equilibrium distribution of a Markov chain. A large number of realizations of the chain are then constructed and the equilibrium distribution obtained from the overall distribution of the state values, typically with an initial section deleted to remove any transient effect of initial conditions.

Although in most cases it will be clear that the chain has reached a stationary state there is the possibility for the chain to be stuck in a sequence of transitions far from equilibrium; this is broadly parallel to the possibility that iterative maximization procedures which appear to have converged in fact have not done so.

One aspect of frequentist discussions concerns the distribution of test statistics, in particular as a basis for finding p-values or confidence limits. The simplest route is to find the maximum likelihood estimate, $\hat{\theta}$, of the full parameter vector and then to sample repeatedly the model $f_Y(y; \hat{\theta})$ and hence find the distribution of concern. For testing a null hypothesis only the nuisance parameter need be estimated. It is only when the conclusions are in some sense borderline that this need be done with any precision. This approach ignores errors in estimating θ. In principle the sampling should be conditional on the observed information $\hat{j}$, the approximately ancillary statistic for the model. An ingenious and powerful alternative available at least in simpler problems is to take the data themselves as forming the population for sampling randomly with replacement. This procedure, called the *(nonparametric) bootstrap*, has improved performance especially if the statistic for study can be taken in pivotal form. The method is related to the notion of *empirical likelihood*, based on an implicit multinomial distribution with support the data points.

6.11 Higher-order asymptotics

A natural development of the earlier discussion is to develop the asymptotic theory to a further stage. Roughly this involves taking series expansions to further terms and thus finding more refined forms of test statistic and establishing distributional approximations better than the normal or chi-squared distributions which underlie the results summarized above. In Bayesian estimation

6.11 Higher-order asymptotics

theory it involves, in particular, using further terms of the Laplace expansion generating approximate normality, in order to obtain more refined approximations to the posterior distribution; alternatively asymptotic techniques more delicate than Laplace approximation may be used and may give substantially improved results.

There are two objectives to such discussions in frequentist theory. The more important aims to give some basis for choosing between the large number of procedures which are equivalent to the first order of asymptotic theory. The second is to provide improved distributional approximations, in particular to provide confidence intervals that have the desired coverage probability to greater accuracy than the results sketched above. In Bayesian theory the second objective is involved.

Typically simple distributional results, such as that the pivot $(\hat{\theta} - \theta)\sqrt{(n\bar{\imath})}$ has asymptotically a standard normal distribution, mean that probability statements about the pivot derived from the standard normal distribution are in error by $O(1/\sqrt{n})$ as n increases. One object of asymptotic theory is thus to modify the statement so that the error is $O(1/n)$ or better. Recall though that the limiting operations are notional and that in any case the objective is to get adequate numerical approximations and the hope would be that with small amounts of data the higher-order results would give better approximations.

The resulting theory is quite complicated and all that will be attempted here is a summary of some conclusions, a number of which underlie some of the recommendations made above.

First it may be possible sometimes to find a Bayesian solution for the posterior distribution in which posterior limits have a confidence limit interpretation to a higher order. Indeed to the first order of asymptotic theory the posterior distribution of the parameter is normal with variance the inverse of the observed information for a wide class of prior distributions, thus producing agreement with the confidence interval solution.

With a single unknown parameter it is possible to find a prior, called a *matching prior*, inducing closer agreement. Note, however, that from conventional Bayesian perspectives this is not an intrinsically sensible objective. The prior serves to import additional information and enforced agreement with a different approach is then irrelevant. While a formal argument will not be given the conclusion is understandable from a combination of the results in Examples 4.10 and 6.3. The location model of Example 4.10 has as its frequentist analysis direct consideration of the likelihood function equivalent to a Bayesian solution with a flat (improper) prior. Example 6.3 shows the transformation of θ that induces constant information, one way of approximating location form. This suggests that a flat prior on the transformed scale, i.e., a prior

density proportional to
$$\sqrt{i(\theta)}, \qquad (6.117)$$
is suitable for producing agreement of posterior interval with confidence interval to a higher order, the prior thus being *second-order matching*. Unfortunately for multidimensional parameters such priors, in general, do not exist.

The simplest general result in higher-order asymptotic theory concerns the profile likelihood statistic W_L. Asymptotically this has a chi-squared distribution with d_ψ degrees of freedom. Were the asymptotic theory to hold exactly we would have
$$E(W_L) = d_\psi. \qquad (6.118)$$
Typically
$$E(W_L) = d_\psi(1 + B/n), \qquad (6.119)$$
where, in general, B is a constant depending on unknown parameters.

This suggests that if we modify W_L to
$$W'_L = W_L/(1 + B/n) \qquad (6.120)$$
the mean of the distribution will be closer to that of the approximating chi-squared distribution. Remarkably, not only is the mean improved but the whole distribution function of the test statistic becomes closer to the approximating form. Any unknown parameters in B can be replaced by estimates. In complicated problems in which theoretical evaluation of B is difficult, simulation of data under the null hypothesis can be used to evaluate B approximately.

The factor $(1 + B/n)$ is called a *Bartlett correction*. When ψ is one-dimensional the two-sided confidence limits for ψ produced from W'_L do indeed have improved coverage properties but unfortunately in most cases the upper and lower limits are not, in general, separately improved.

Notes 6

Section 6.2. A proof of the asymptotic results developed informally here requires regularity conditions ensuring first that the root of the maximum likelihood estimating equation with the largest log likelihood converges in probability to θ. Next conditions are needed to justify ignoring the third-order terms in the series expansions involved. Finally the Central Limit Theorem for the score and the convergence in probability of the ratio of observed to expected information have to be proved. See van der Vaart (1998) for an elegant and mathematically careful account largely dealing with independent and identically distributed random variables. As has been mentioned a number of times in the text, a key issue in applying these ideas is whether, in particular, the distributional results are an adequate approximation in the particular application under study.

A standard notation in analysis is to write $\{a_n\} = O(n^b)$ as $n \to \infty$ to mean that a_n/n^b is bounded and to write $a_n = o(n^b)$ to mean $a_n/n^b \to 0$. The corresponding notation in probability theory is to write for random variables A_n that $A_n = O_p(n^b)$ or $A_n = o_p(n^b)$ to mean respectively that A_n/n^b is bounded with very high probability and that it tends to zero in probability respectively. Sometimes one says in the former case that a_n or A_n is of the order of magnitude n^b. This is a use of the term different from that in the physical sciences where often an order of magnitude is a power of 10. If $A_n - a = o_p(1)$ then A_n *converges in probability* to a.

In discussing distributional approximations, in particular normal approximations, the terminology is that V_n is asymptotically normal with mean and variance respectively μ_n and σ_n^2 if the distribution function of $(V_n - \mu_n)/\sigma_n$ tends to the distribution function of the standard normal distribution. Note that μ_n and σ_n refer to the approximating normal distribution and are not necessarily approximations to the mean and variance of V_n.

The relation and apparent discrepancy between Bayesian and frequentist tests was outlined by Jeffreys (1961, first edition 1939) and developed in detail by Lindley (1957). A resolution broadly along the lines given here was sketched by Bartlett (1957); see, also, Cox and Hinkley (1974).

Section 6.3. Another interpretation of the transformation results for the score and the information matrix is that they play the roles respectively of a contravariant vector and a second-order covariant tensor and the latter could be taken as the metric tensor in a Riemannian geometry. Most effort on geometrical aspects of asymptotic theory has emphasized the more modern coordinate-free approach. See Dawid (1975), Amari (1985) and Murray and Rice (1993). For a broad review, see Barndorff-Nielsen *et al.* (1986) and for a development of the Riemannian approach McCullagh and Cox (1986).

Section 6.5. For more on Bayes factors, see Kass and Raftery (1995) and Johnson (2005).

Section 6.8. Asymptotic tests based strongly on the method of Lagrange multipliers are developed systematically by Aitchison and Silvey (1958) and Silvey (1959). The discussion here is based on joint work with N. Wermuth and

G. Marchetti. Covariance selection models involving zero constraints in concentration matrices were introduced by Dempster (1972). Models with zero constraints on covariances are a special class of linear in covariance structures (Anderson, 1973). Both are of current interest in connection with the study of various kinds of Markov independence graphs.

The discussion of a covering model as a systematic device for generating close approximations to maximum likelihood estimates follows Cox and Wermuth (1990); R. A. Fisher noted that one step of an iterative scheme for maximum likelihood estimates recovered all information in a large-sample sense.

Section 6.10. Nelder and Mead (1965) give a widely used algorithm based on simplex search. Lange (2000) gives a general account of numerical-analytic problems in statistics. For the EM algorithm see Dempster *et al.* (1977) and for a review of developments Meng and van Dyk (1997). Sundberg (1974) develops formulae for incomplete data from an exponential family model. Efron (1979) introduces the nonparametric bootstrap; see Davison and Hinkley (1997) for a very thorough account. Markov chain Monte Carlo methods are discussed from somewhat different perspectives by Liu (2002) and by Robert and Casella (2004). For a wide-ranging account of simulation methods, see Ripley (1987). For an example of the envelope method of computing a profile likelihood, see Cox and Medley (1989).

Section 6.11. Higher-order asymptotic theory of likelihood-based statistics was initially studied by M. S. Bartlett beginning with a modification of the likelihood ratio statistic (Bartlett, 1937) followed by the score statistic (Bartlett, 1953a, 1953b). Early work explicitly or implicitly used Edgeworth expansions, modifications of the normal distribution in terms of orthogonal polynomials derived via Taylor expansion of a moment generating function. Nearly 30 years elapsed between the introduction (Daniels, 1954) of saddle-point or tilted Edgeworth series and their systematic use in inferential problems. Butler (2007) gives a comprehensive account of the saddle-point method and its applications. Efron and Hinkley (1978) establish the superiority of observed over expected information for inference by appeal to conditionality arguments. For a general account, see Barndorff-Nielsen and Cox (1994) and for a wide-ranging more recent review Reid (2003). The book by Brazzale *et al.* (2007) deals both with the theory and the implementation of these ideas. This latter work is in a likelihood-based framework. It may be contrasted with a long series of papers collected in book form by Akahira and Takeuchi (2003) that are very directly in a Neyman–Pearson setting.

7
Further aspects of maximum likelihood

Summary. Maximum likelihood estimation and related procedures provide effective solutions for a wide range of problems. There can, however, be difficulties leading at worst to inappropriate procedures with properties far from those sketched above. Some of the difficulties are in a sense mathematical pathologies but others have serious statistical implications. The first part of the chapter reviews the main possibilities for anomalous behaviour. For illustration relatively simple examples are used, often with a single unknown parameter. The second part of the chapter describes some modifications of the likelihood function that sometimes allow escape from these difficulties.

7.1 Multimodal likelihoods

In some limited cases, notably connected with exponential families, convexity arguments can be used to show that the log likelihood has a unique maximum. More commonly, however, there is at least the possibility of multiple maxima and saddle-points in the log likelihood surface. See Note 7.1.

There are a number of implications. First, proofs of the convergence of algorithms are of limited comfort in that convergence to a maximum that is in actuality not the overall maximum of the likelihood is unhelpful or worse. Convergence to the global maximum is nearly always required for correct interpretation. When there are two or more local maxima giving similar values to the log likelihood, it will in principle be desirable to know them all; the natural confidence set may consist of disjoint intervals surrounding these local maxima. This is probably an unusual situation, however, and the more common situation is that the global maximum is dominant.

The discussion of Sections 6.4–6.7 of the properties of profile likelihood hinges only on the estimate of the nuisance parameter $\hat{\lambda}_\psi$ being a stationary point of the log likelihood given ψ and on the formal information matrix $j_{\psi\psi\cdot\lambda}$ being positive definite. There is thus the disturbing possibility that the profile log likelihood appears well-behaved but that it is based on defective estimates of the nuisance parameters.

The argument of Section 6.8 can sometimes be used to obviate some of these difficulties. It starts from a set of estimates with simple well-understood properties, corresponding to a saturated model, for example. The required maximum likelihood estimates under the restricted model are then expressed as small corrections to some initial simple estimates. Typically the corrections being $O_p(1/\sqrt{n})$ in magnitude should be of the order of the standard error of the initial estimates, and small compared with that if the initial estimates are of high efficiency. There are two consequences. If the correction is small, then it is quite likely that close to the global maximum of the likelihood has been achieved. If the correction is large, the proposed model is probably a bad fit to the data. This rather imprecise argument suggests that anomalous forms of log likelihood are likely to be of most concern when the model is a bad fit to the data or when the amount of information in the data is very small.

It is possible for the log likelihood to be singular, i.e., to become unbounded in a potentially misleading way. We give the simplest example of this.

Example 7.1. *An unbounded likelihood.* Suppose that $Y_1, \ldots, Y_n$ are independently and identically distributed with unknown mean μ and density

$$(1 - b)\phi(y - \mu) + b\lambda^{-1}\phi\{(y - \mu)/\lambda\}, \tag{7.1}$$

where $\phi(.)$ is the standard normal density, λ is a nuisance parameter and b is a known small constant, for example 10^{-6}. Because for most combinations of the parameters the second term in the density is negligible, the log likelihood will for many values of λ have a maximum close to $\mu = \bar{y}$, and $\bar{y}$, considered as an estimate of μ, will enjoy the usual properties. But for $\mu = y_k$ for any k the log likelihood will be unbounded as $\lambda \to 0$. Essentially a discrete atom, the limiting form, gives an indefinitely larger log likelihood than any continuous distribution. To some extent the example is pathological. A partial escape route at a fundamental level can be provided as follows. All data are discrete and the density is used to give approximations for small bin widths (grouping intervals) h. For any value of μ for which an observation is reasonably central to a bin and for which λ is small compared with h one normal component in the model attaches a probability of 1 to that observation and hence the model attaches likelihood b, namely the weight of the normal component in question. Hence the ratio of the likelihood achieved at $\mu = y_k$ and small λ to that achieved at

$\mu = \bar{y}$ with any λ not too small is

$$\frac{b\sqrt{(2\pi)}\exp\{-n(y_k - \bar{y})^2/2\}}{h(1-b)}. \quad (7.2)$$

This ratio is of interest because it has to be greater than 1 for the optimum likelihood around $\mu = y_k$ to dominate the likelihood at $\mu = \bar{y}$. Unless b/h is large and y_k is close to $\bar{y}$ the ratio will be less than 1. The possibility of anomalous behaviour with misleading numerical consequences is much more serious with the continuous version of the likelihood.

7.2 Irregular form

A key step in the discussion of even the simplest properties is that the maximum likelihood estimating equation is unbiased, i.e., that

$$E\{\nabla l(\theta; Y); \theta\} = 0. \quad (7.3)$$

This was proved by differentiating under the integral sign the normalizing equation

$$\int f(y; \theta) dy = 1. \quad (7.4)$$

If this differentiation is not valid there is an immediate difficulty and we may call the problem *irregular*. All the properties of maximum likelihood estimates sketched above are now suspect. The most common reason for failure is that the range of integration effectively depends on θ.

The simplest example of this is provided by a uniform distribution with unknown range.

Example 7.2. *Uniform distribution.* Suppose that $Y_1, \ldots, Y_n$ are independently distributed with a uniform distribution on $(0, \theta)$, i.e., have constant density $1/\theta$ in that range and zero elsewhere. The normalizing condition for n independent observations is, integrating in R^n,

$$\int_0^\theta (1/\theta^n) dy = 1. \quad (7.5)$$

If this is differentiated with respect to θ there is a contribution not only from differentiating $1/\theta$ but also one from the upper limit of the integral and the key property fails.

Direct examination of the likelihood shows it to be discontinuous, being zero if $\theta < y_{(n)}$, the largest observation, and equal to $1/\theta^n$ otherwise. Thus $y_{(n)}$ is the maximum likelihood estimate and is also the minimal sufficient statistic for θ. Its properties can be studied directly by noting that

$$P(Y_{(n)} \leq z) = \Pi P(Y_j \leq z) = (z/\theta)^n. \tag{7.6}$$

For inference about θ, consider the pivot $n(\theta - Y_{(n)})/\theta$ for which

$$P\{n(\theta - Y_{(n)})/\theta < t\} = (1 - t/n)^n, \tag{7.7}$$

so that as n increases the pivot is asymptotically exponentially distributed. In this situation the maximum likelihood estimate is a sure lower limit for θ, the asymptotic distribution is not Gaussian and, particularly significantly, the errors of estimation are $O_p(1/n)$ not $O_p(1/\sqrt{n})$. An upper confidence limit for θ is easily calculated from the exact or from the asymptotic pivotal distribution.

Insight into the general problem can be obtained from the following simple generalization of Example 7.2.

Example 7.3. *Densities with power-law contact.* Suppose that $Y_1, \ldots, Y_n$ are independently distributed with the density $(a + 1)(\theta - y)^a/\theta^{a+1}$, where a is a known constant. For Example 7.2, $a = 0$. As a varies the behaviour of the density near the critical end-point θ changes.

The likelihood has a change of behaviour, although not necessarily a local maximum, at the largest observation. A slight generalization of the argument used above shows that the pivot

$$n^{1/(a+1)}(\theta - Y_{(n)})/\theta \tag{7.8}$$

has a limiting Weibull distribution with distribution function $1 - \exp(-t^{a+1})$. Thus for $-1/2 < a < 1$ inference about θ is possible from the maximum observation at an asymptotic rate faster than $1/\sqrt{n}$.

Consider next the formal properties of the maximum likelihood estimating equation. We differentiate the normalizing equation

$$\int_0^\theta \{(a+1)(\theta - y)^a/\theta^{a+1}\}dy = 1 \tag{7.9}$$

obtaining from the upper limit a contribution equal to the argument of the integrand there and thus zero if $a > 0$ and nonzero (and possibly unbounded) if $a \leq 0$.

For $0 < a \leq 1$ the score statistic for a single observation is

$$\partial \log f(Y, \theta)/\partial \theta = a/(\theta - Y) - (a+1)/\theta \tag{7.10}$$

and has zero mean and infinite variance; an estimate based on the largest observation is preferable.

7.2 Irregular form

For $a > 1$ the normalizing condition for the density can be differentiated twice under the integral sign, Fisher's identity relating the expected information to the variance of the score holds and the problem is regular.

The relevance of the extreme observation for asymptotic inference thus depends on the level of contact of the density with the y-axis at the terminal point. This conclusion applies also to more complicated problems with more parameters. An example is the displaced exponential distribution with density

$$\rho \exp\{-\rho(y - \theta)\} \tag{7.11}$$

for $y > \theta$ and zero otherwise, which has a discontinuity in density at $y = \theta$. The sufficient statistics in this case are the smallest observation $y_{(1)}$ and the mean. Estimation of θ by $y_{(1)}$ has error $O_p(1/n)$ and estimation of ρ with error $O_p(1/\sqrt{n})$ can proceed as if θ is known at least to the first order of asymptotic theory. The estimation of the displaced Weibull distribution with density

$$\rho\gamma\{\rho(y - \theta)\}^{\gamma - 1} \exp[-\{\rho(y - \theta)\}^\gamma] \tag{7.12}$$

for $y > \theta$ illustrates the richer possibilities of Example 7.3.

There is an important additional aspect about the application of these and similar results already discussed in a slightly different context in Example 7.1. The observations are treated as continuously distributed, whereas all real data are essentially recorded on a discrete scale, often in groups or bins of a particular width. For the asymptotic results to be used to obtain approximations to the distribution of extremes, it is, as in the previous example, important that the grouping interval in the relevant range is chosen so that rounding errors are unimportant relative to the intrinsic random variability of the continuous variables in the region of concern. If this is not the case, it may be more relevant to consider an asymptotic argument in which the grouping interval is fixed as n increases. Then with high probability the grouping interval containing the terminal point of support will be identified and estimation of the position of the maximum within that interval can be shown to be an essentially regular problem leading to a standard error that is $O(1/\sqrt{n})$.

While the analytical requirements justifying the expansions of the log likelihood underlying the theory of Chapter 6 are mild, they can fail in ways less extreme than those illustrated in Examples 7.2 and 7.3. For example the log likelihood corresponding to n independent observations from the Laplace density $\exp(-|y - \theta|)/2$ leads to a score function of

$$-\Sigma \text{sgn}(y_k - \theta). \tag{7.13}$$

It follows that the maximum likelihood estimate is the median, although if n is odd the score is not differentiable there. The second derivative of the log likelihood is not defined, but it can be shown that the variance of the score does determine the asymptotic variance of the maximum likelihood estimate.

A more subtle example of failure is provided by a simple time series problem.

Example 7.4. *Model of hidden periodicity.* Suppose that $Y_1, \ldots, Y_n$ are independently normally distributed with variance σ^2 and with

$$E(Y_k) = \mu + \alpha \cos(k\omega) + \beta \sin(k\omega), \tag{7.14}$$

where $(\mu, \alpha, \beta, \omega, \sigma^2)$ are unknown. A natural interpretation is to think of the observations as equally spaced in time. It is a helpful simplification to restrict ω to the values

$$\omega_p = 2\pi p/n \quad (p = 1, 2, \ldots, [n/2]), \tag{7.15}$$

where $[x]$ denotes the integer part of x. We denote the true value of ω holding in (7.14) by ω_q.

The finite Fourier transform of the data is defined by

$$\tilde{Y}(\omega_p) = \sqrt{(2/n)} \Sigma Y_k e^{ik\omega_p}$$
$$= \tilde{A}(\omega_p) + i\tilde{B}(\omega_p). \tag{7.16}$$

Because of the special properties of the sequence $\{\omega_p\}$ the transformation from $Y_1, \ldots, Y_n$ to $\Sigma Y_k/\sqrt{n}, \tilde{A}(\omega_1), \tilde{B}(\omega_1), \ldots$ is orthogonal and hence the transformed variables are independently normally distributed with variance σ^2. Further, if the linear model (7.14) holds with $\omega = \omega_q$ known, it follows that:

- for $p \neq q$, the finite Fourier transform has expectation zero;
- the least squares estimates of α and β are $\sqrt{(2/n)}$ times the real and imaginary parts of $\tilde{Y}(\omega_q)$, the expectations of which are therefore $\sqrt{(n/2)}\alpha$ and $\sqrt{(n/2)}\beta$;
- the residual sum of squares is the sum of squares of all finite Fourier transform components except those at ω_q.

Now suppose that ω is unknown and consider its profile likelihood after estimating $(\mu, \alpha, \beta, \sigma^2)$. Equivalently, because (7.14) for fixed ω is a normal-theory linear model, we may use minus the residual sum of squares. From the points listed above it follows that the residual sum of squares varies randomly across the values of ω except at $\omega = \omega_q$, where the residual sum of squares is much smaller. The dip in values, corresponding to a peak in the profile likelihood, is isolated at one point. Even though the log likelihood is differentiable as a function of the continous variable ω the fluctuations in its value are so rapid and extreme that two terms of a Taylor expansion about the maximum likelihood point are totally inappropriate.

In a formal asymptotic argument in which (α, β) are $O(1/\sqrt{n})$, and so on the borderline of detectability, the asymptotic distribution of $\hat{\omega}$ is a mixture of an atom at the true value ω_q and a uniform distribution on $(0, \pi/2)$. The uniform distribution arises from identifying the maximum at the wrong point.

7.3 Singular information matrix

A quite different kind of anomalous behaviour occurs if the score statistic is defined but is singular. In particular, for a scalar parameter the score function may be identically zero for special values of the parameter. It can be shown that this can happen only on rather exceptional sets in the parameter space but if one of these points corresponds to a null hypothesis of interest, the singularity is of statistical concern; see Note 7.3.

Example 7.5. *A special nonlinear regression.* Suppose that $Y_1, \ldots, Y_n$ are independently normally distributed with unit variance and that for constants $z_1, \ldots, z_n$ not all equal we have one of the models

$$E(Y_j) = \exp(\theta z_j) - 1 - \theta z_j \tag{7.17}$$

or

$$E(Y_j) = \exp(\theta z_j) - 1 - \theta z_j - (\theta z_j)^2/2. \tag{7.18}$$

Consider the null hypothesis $\theta = 0$. For the former model the contribution to the score statistic from Y_j is

$$-(\partial/\partial\theta)\{Y_j - \exp(\theta z_j) + 1 + \theta z_j\}^2/2 \tag{7.19}$$

and both this and the corresponding contribution to the information are identically zero at $\theta = 0$.

It is clear informally that, to extend the initial discussion of maximum likelihood estimation and testing to such situations, the previous series expansions leading to likelihood derivatives must be continued to enough terms to obtain nondegenerate results. Details will not be given here. Note, however, that locally near $\theta = 0$ the two forms for $E(Y_j)$ set out above are essentially $(\theta z_j)^2/2$ and $(\theta z_j)^3/6$. This suggests that in the former case θ^2 will be estimated to a certain precision, but that the sign of θ will be estimated with lower precision, whereas that situation will not arise in the second model θ^3, and hence the sign of θ, being estimable.

A more realistic example of the same issue is provided by simple models of informative nonresponse.

Example 7.6. *Informative nonresponse.* Suppose that $Y_1, \ldots, Y_n$ are independently normally distributed with mean μ and variance σ^2. For each observation there is a possibility that the corresponding value of y cannot be observed. Note that in this formulation the individual study object is always observed, it being only the value of Y that may be missing. If the probability of being unobserved is independent of the value of y the analysis proceeds with the likelihood of the observed values; the values of y are missing completely at random. Suppose, however, that given $Y = y$ the probability that y is observed has the form

$$\exp\{H(\alpha_0 + \alpha_1(y - \mu)/\sigma)\}, \tag{7.20}$$

where α_0, α_1 are unknown parameters and $H(.)$ is a known function, for example corresponding to a logistic regression of the probability of being observed as a function of y. Missing completely at random corresponds to $\alpha_1 = 0$ and interest may well focus on that null hypothesis and more generally on small values of α_1.

We therefore consider two random variables (R, Y), where R is binary, taking values 0 and 1. The value of Y is observed if and only if $R = 1$. The contribution of a single individual to the log likelihood is thus

$$-r \log \sigma - r(y - \mu)^2/(2\sigma^2) + rH\{\alpha_0 + \alpha_1(y - \mu)/\sigma\}$$
$$+ (1 - r) \log E[1 - \exp\{H(\alpha_0 + \alpha_1(Y - \mu)/\sigma)\}]. \tag{7.21}$$

The last term is the overall probability that Y is not observed. The primary objective in this formulation would typically be the estimation of μ allowing for the selectively missing data values; in a more realistic setting the target would be a regression coefficient assessing the dependence of Y on explanatory variables.

Expansion of the log likelihood in powers of α_1 shows that the log likelihood is determined by

$$n_c, \; \Sigma_c(y_k - \mu), \; \Sigma_c(y_k - \mu)^2, \tag{7.22}$$

and so on, where n_c is the number of complete observations and Σ_c denotes sum over the complete observations. That is, the relevant functions of the data are the proportion of complete observations and the first few sample moments of the complete observations.

If the complete system variance σ^2 is assumed known, estimation may be based on n_c and the first two moments, thereby estimating the three unknown parameters $(\mu, \alpha_0, \alpha_1)$. If σ^2 is regarded as unknown, then the sample third moment is needed also.

The general difficulty with this situation can be seen in various ways. One is that when, for example, σ^2 is assumed known, information about the missing structure can be inferred by comparing the observed sample variance with the known value and ascribing discrepancies to the missingness. Not only will such comparisons be of low precision but more critically they are very sensitive to the accuracy of the presumed value of variance. When the variance is unknown the comparison of interest involves the third moment. Any evidence of a nonvanishing third moment is ascribed to the missingness and this makes the inference very sensitive to the assumed symmetry of the underlying distribution.

More technically the information matrix is singular at the point $\alpha_1 = 0$. Detailed calculations not given here show a parallel with the simpler regression problem discussed above. For example when σ^2 is unknown, expansion yields the approximate estimating equation

$$\tilde{\alpha}_1^3 = \{H'''(\tilde{\alpha}_0)\}^{-1}\hat{\gamma}_3, \tag{7.23}$$

where $\hat{\gamma}_3$ is the standardized third cumulant, i.e., third cumulant or moment about the mean divided by the cube of the standard deviation. It follows that fluctuations in $\tilde{\alpha}_1$ are $O_p(n^{-1/6})$; even apart from this the procedure is fragile for the reason given above.

7.4 Failure of model

An important possibility is that the data under analysis are derived from a probability density $g(.)$ that is not a member of the family $f(y;\theta)$ originally chosen to specify the model. Note that since all models are idealizations the empirical content of this possibility is that the data may be seriously inconsistent with the assumed model and that although a different model is to be preferred, it is fruitful to examine the consequences for the fitting of the original family.

We assume that for the given g there is a value of θ in the relevant parameter space such that

$$E_g\{\nabla l(\theta)\}_{\theta=\theta_g} = 0. \tag{7.24}$$

Here E_g denotes expectation with respect to g.

The value θ_g minimizes

$$\int g(y)\log\{f_Y(y;\theta)/g(y)\}dy. \tag{7.25}$$

This, the Kullback–Leibler divergence, arose earlier in Section 5.9 in connection with reference priors.

A development of (7.24) is to show that the asymptotic distribution of $\hat{\theta}$ under the model g is asymptotically normal with mean θ_g and covariance matrix

$$E_g\{\nabla\nabla^T \log f_Y(y;\theta)\} \text{cov}_g\{\nabla \log f_Y(y;\theta)\}\{E_g\{\nabla\nabla^T \log f_Y(y;\theta)\}\}^T, \quad (7.26)$$

which is sometimes called the *sandwich formula*.

The most immediate application of these formulae is to develop likelihood ratio tests for what are sometimes called separate families of hypotheses. Suppose that we regard consistency with the family $f_Y(y;\theta)$ as a null hypothesis and either a single distribution $g_Y(y)$ or more realistically a family $g_Y(y;\omega)$ as the alternative. It is assumed that for a given θ there is no value of ω that exactly agrees with $f_Y(y;\theta)$ but that in some sense there are values for which there is enough similarity that discrimination between them is difficult with the data under analysis. The likelihood ratio statistic

$$l_g(\hat{\omega}) - l_f(\hat{\theta}) \quad (7.27)$$

can take negative as well as positive values, unlike the case where the null hypothesis model is nested within the alternatives. It is clear therefore that the asymptotic distribution cannot be of the chi-squared form. In fact, it is normal with a mean and variance that can be calculated via (7.26). Numerical work shows that approach to the limiting distribution is often very slow so that a more sensible course in applications will often be to compute the limiting distribution by simulation. In principle the distribution should be found conditionally on sufficient statistics for the null model and at least approximately these are $\hat{\theta}$ and the corresponding observed information. Usually it will be desirable to repeat the calculation interchanging the roles of the two models, thereby assessing whether both, one but not the other, or neither model is adequate in the respect considered.

Sometimes more refined calculations are possible, notably when one or preferably both models have simple sufficient statistics.

7.5 Unusual parameter space

In many problems the parameter space of possible values is an open set in some d-dimensional space. That is, the possibility of parameter points on the boundary of the space is not of particular interest. For example, probabilities are constrained to be in $(0, 1)$ and the possibility of probabilities being exactly 0 or 1 is usually not of special moment. Parameter spaces in applications are commonly formulated as infinite although often in practice there are limits on the values that can arise.

7.5 Unusual parameter space

Constraints imposed on the parameter space may be highly specific to a particular application. We discuss a simple example which is easily generalized.

Example 7.7. *Integer normal mean.* Suppose that $\bar{Y}$ is the mean of n independent random variables normally distributed with unknown mean μ and known variance σ_0^2. Suppose that the unknown mean μ is constrained to be a nonnegative integer; it might, for example, be the number of carbon atoms in some moderately complicated molecule, determined indirectly with error. It is simplest to proceed via the connection with significance tests. We test in turn consistency with each nonnegative integer r, for example by a two-sided test, and take all those r consistent with the observed value $\bar{y}$ as a confidence set. That is, at level $1 - 2c$ the set of integers in the interval

$$(\bar{y} - k_c^* \sigma_0/\sqrt{n}, \bar{y} + k_c^* \sigma_0/\sqrt{n}) \tag{7.28}$$

forms a confidence set. That is, any true integer value of μ is included with appropriate probability. Now if $\sigma_0/\sqrt{n}$ is small compared with 1 there is some possibility that the above set is null, i.e., no integer value is consistent with the data at the level in question.

From the perspective of significance testing, this possibility is entirely reasonable. To achieve consistency between the data and the model we would have to go to more extreme values of c and if that meant very extreme values then inconsistency between data and model would be signalled. That is, a test of model adequacy is automatically built into the procedure. From a different interpretation of confidence sets as sets associated with data a specified proportion of which are correct, there is the difficulty that null sets are certainly false, if the model is correct. This is a situation in which the significance-testing formulation is more constructive in leading to sensible interpretations of data.

These particular formal issues do not arise in a Bayesian treatment. If the prior probability that $\mu = r$ is π_r, then the corresponding posterior probability is

$$\pi_r \exp(r\bar{y}\gamma - r^2\gamma/2) \Big/ \sum_{s=1}^{\infty} \pi_s \exp(s\bar{y}\gamma - s^2\gamma/2), \tag{7.29}$$

where $\gamma = n/\sigma_0^2$. When γ is large the posterior probability will, unless the fractional part of $\bar{y}$ is very close to 1/2, concentrate strongly on one value, even though that value may, in a sense, be strongly inconsistent with the data. Although the Bayesian solution is entirely sensible in the context of the formulation adopted, the significance-testing approach seems the more fruitful in terms of learning from data. In principle the Bayesian approach could, of course, be extended to allow some probability on noninteger values.

There are more complicated possibilities, for example when the natural parameter space is in mathematical terms not a manifold.

Example 7.8. *Mixture of two normal distributions.* Suppose that it is required to fit a mixture of two normal distributions, supposed for simplicity to have known and equal variances. There are then three parameters, two means μ_1, μ_2 and an unknown mixing probability π. Some interest might well lie in the possibility that one component is enough. This hypothesis can be represented as $\pi = 0, \mu_1$ arbitrary, or as $\pi = 1, \mu_2$ arbitrary, or again as $\mu_1 = \mu_2$, this time with π arbitrary. The standard asymptotic distribution theory of the likelihood ratio statistics does not apply in this case.

The general issue here is that under the null hypothesis parameters defined under the general family become undefined and therefore not estimable. One formal possibility is to fix the unidentified parameter at an arbitrary level and to construct the corresponding score statistic for the parameter of interest, then maximize over the arbitrary level of the unidentified parameter. This argument suggests that the distribution of the test statistic under the null hypothesis is that of the maximum of a Gaussian process. While the distribution theory can be extended to cover such situations, the most satisfactory solution is probably to use the likelihood ratio test criterion and to obtain its null hypothesis distribution by simulation.

If the different components have different unknown variance, the apparent singularities discussed in Example 7.1 recur.

7.6 Modified likelihoods

7.6.1 Preliminaries

We now consider methods based not on the log likelihood itself but on some modification of it. There are three broad reasons for considering some such modification.

- In models in which the dimensionality of the nuisance parameter is large direct use of maximum likelihood estimates may be very unsatisfactory and then some modification of either the likelihood or the method of estimation is unavoidable.
- Especially in problems with non-Markovian dependency structure, it may be difficult or impossible to compute the likelihood in useful form. Some time series and many spatial and spatial-temporal models are like that.
- The likelihood may be very sensitive to aspects of the model that are thought particularly unrealistic.

7.6.2 Simple example with many parameters

We begin by discussing the first possibility as the most important, although we shall give examples of the other two later.

We give an example where direct use of maximum likelihood estimation is misleading. That it is a type of model which the exact theory in the first part of the book can handle makes comparative discussion easier.

Example 7.9. *Normal-theory linear model with many parameters.* Suppose that the $n \times 1$ random vector Y has components that are independently normally distributed with variance τ and that $E(Y) = z\beta$, where z is an $n \times d$ matrix of full rank $d < n$ and β is a $d \times 1$ vector of unknown parameters. The log likelihood is

$$-(n/2)\log \tau - (Y - z\beta)^T(Y - z\beta)/(2\tau). \tag{7.30}$$

Now suppose that τ is the parameter of interest. The maximum likelihood estimate of β is, for all τ, the least squares estimate $\hat{\beta}$. It follows that

$$\hat{\tau} = (Y - z\hat{\beta})^T(Y - z\hat{\beta})/n, \tag{7.31}$$

the residual sum of squares divided by sample size, not by $n - d$, the degrees of freedom for residual. Direct calculation shows that the expected value of the residual sum of squares is $(n - d)\tau$, so that

$$E(\hat{\tau}) = (1 - d/n)\tau. \tag{7.32}$$

Conventional asymptotic arguments take d to be fixed as n increases; the factor $(1 - d/n)$ tends to 1 and it can be verified that all the standard properties of maximum likelihood estimates hold. If, however, d is comparable with n, then $\hat{\tau}$ is for large n very likely to be appreciably less than τ; systematic overfitting has taken place.

Two quite realistic special cases are first factorial experiments in which the number of main effects and interactions fitted may be such as to leave relatively few degrees of freedom available for estimating τ. A second illustration is a matched pair model in which Y is formed from m pairs (Y_{k0}, Y_{k1}) with

$$E(Y_{k0}) = \lambda_k - \Delta, E(Y_{k1}) = \lambda_k + \Delta. \tag{7.33}$$

Here $n = 2m$ and $d = m + 1$ so that $\hat{\tau}$ underestimates τ by a factor of 2. There is the special feature in this case that the λ_k are *incidental parameters*, meaning that each only affects a limited number of observations, in fact two. Thus the supposition underlying maximum likelihood theory that all errors of estimation are small is clearly violated. This is the reason in this instance that direct use of maximum likelihood is so unsatisfactory.

The same explanation does not apply so directly to the factorial design where, with a suitable parameterization in terms of main effects and interactions of various orders, all regression parameters representing these contrasts can be estimated with high precision. The same issue is, however, latent within this example in that we could, somewhat perversely, reparameterize into a form in which the matrix of coefficients defining the linear model is sparse, i.e., contains many zeros, and then the same point could arise.

While appeal to asymptotic maximum likelihood theory is not needed in the previous problem, it provides a test case. If maximum likelihood theory is to be applied with any confidence to similar more complicated problems an escape route must be provided. There are broadly three such routes. The first is to avoid, if at all possible, formulations with many nuisance parameters; note, however, that this approach would rule out semiparametric formulations as even more extreme. The second, and somewhat related, approach is applicable to problems where many or all of the individual parameters are likely to take similar values and have similar interpretations. This is to use a random effects formulation in which, in the matched pair illustration, the representation

$$Y_{k0} = \mu - \Delta + \eta_k + \epsilon_{k0}, \quad Y_{k1} = \mu + \Delta + \eta_k + \epsilon_{k1} \tag{7.34}$$

is used. Here the η_k and the ϵ_{ks} are mutually independent normal random variables of zero mean and variances respectively τ_η, τ_ϵ and where τ_ϵ, previously denoted by τ, is the parameter of interest.

Example 7.10. *A non-normal illustration.* In generalization of the previous example, and to show that the issues are in no way specific to the normal distribution, suppose that (Y_{l0}, Y_{l1}) are for $l = 1, \ldots, m$ independently distributed in the exponential family form

$$m(y) \exp\{y\phi - k(\phi)\} \tag{7.35}$$

with canonical parameters (ϕ_{l0}, ϕ_{l1}) such that

$$\phi_{l1} = \lambda_l + \psi, \quad \phi_{l0} = \lambda_l - \psi. \tag{7.36}$$

Thus the log likelihood is

$$\Sigma t_l \lambda_l + \psi \Sigma d_l - \Sigma\{k(\lambda_l + \psi) + k(\lambda_l - \psi)\}, \tag{7.37}$$

where $t_l = y_{l1} + y_{l0}, d_l = y_{l1} - y_{l0}$.

The formal maximum likelihood equations are thus

$$t_l = k'(\hat{\lambda}_l + \hat{\psi}) + k'(\hat{\lambda}_l - \hat{\psi}), \quad \Sigma d_l = \Sigma k'(\hat{\lambda}_l + \hat{\psi}) - k'(\hat{\lambda}_l - \hat{\psi}).$$

$$\tag{7.38}$$

If we argue locally near $\hat{\psi} = 0$ we have that approximately

$$\hat{\psi} = \frac{\Sigma d_l}{2\Sigma k''(\hat{\lambda}_l)}, \tag{7.39}$$

the limit in probability of which is

$$\psi \frac{\lim n^{-1}\Sigma k''(\lambda_l)}{\lim_p n^{-1}\Sigma k''(\hat{\lambda}_l)}. \tag{7.40}$$

In general the factor multiplying ψ is not 1. If, however, $k''(\lambda)$ is constant or linear, i.e, $k(.)$ is quadratic or cubic, the ratio is 1. The former is the special case of a normal distribution with known variance and unknown mean and in this case the difference in means is the limit in probability of the maximum likelihood estimate, as is easily verified directly.

Another special case is the Poisson distribution of mean μ and canonical parameter $\phi = \log \mu$, so that the additive model in ϕ is a multiplicative representation for the mean. Here $k''(\phi) = k(\phi) = e^\phi = \mu$. It follows again that the ratio in question tends to 1 in probability. It is likely that in all other cases the maximum likelihood estimate of ψ is inconsistent.

7.6.3 Requirements for a modified likelihood

The final route to resolving these difficulties with direct use of maximum likelihood, which we now develop, involves modifying the likelihood function.

Suppose that the maximum likelihood estimating equation

$$[\nabla_\theta l(\theta;y)]_{\hat{\theta}} = \nabla l(\hat{\theta};y) = 0 \tag{7.41}$$

is replaced for some suitable function q by

$$\nabla q(\tilde{\theta};y) = 0. \tag{7.42}$$

In order for the main properties of maximum likelihood estimates to apply preferably in some extended circumstances, we require that:

- the estimating equation is unbiased in the sense that

$$E\{\nabla q(\theta;Y);\theta\} = 0; \tag{7.43}$$

- the analogue of Fisher's identity that relates the variance of the score to the information holds, namely that

$$E\{\nabla \nabla^T q(\theta;Y) + \nabla q(\theta;Y)\nabla^T q(\theta;Y)\} = 0; \tag{7.44}$$

- an asymptotic dependence on n obtains similar to that for maximum likelihood estimates. Then the asymptotic covariance matrix of $\tilde{\theta}$ is given by the inverse of a matrix analogous to an information matrix;

- preferably in some sense all or most of the available information is recovered.

We call the first two properties just given first- and second-order validity respectively.

First-order validity is essential for the following discussion to apply. If first- but not second-order validity holds the asymptotic covariance matrix is given by the slightly more complicated sandwich formula of Section 7.4.

We now give a number of examples of modified likelihoods to which the above discussion applies.

7.6.4 Examples of modified likelihood

We begin by returning to Example 7.9, the general linear normal-theory model. There are two approaches to what amounts to the elimination from the problem of the nuisance parameters, in this case the regression parameter β.

We can find an orthogonal matrix, b, such that the first d rows are in the space spanned by the columns of z. Then the transformed variables $V = bY$ are such that $E(V_{d+1}) = 0, \ldots, E(V_n) = 0$. Moreover, because of the orthogonality of the transformation, the V_s are independently normally distributed with constant variance τ. The least squares estimates $\hat{\beta}$ are a nonsingular transformation of $v_1, \ldots, v_d$ and the residual sum of squares is RSS $= v_{d+1}^2 + \cdots + v_n^2$.

Therefore the likelihood is the product of a factor from $V_1, \ldots, V_d$, or equivalently from the least squares estimates $\hat{\beta}$, and a factor from the remaining variables $V_{d+1}, \ldots, V_n$. That is, the likelihood is

$$L_1(\beta, \tau; \hat{\beta}) L_2(\tau; v_{d+1}, \ldots, v_n). \tag{7.45}$$

Suppose we discard the factor L_1 and treat L_2 as the likelihood for inference about τ. There is now a single unknown parameter and, the sample size being $n - d$, the maximum likelihood estimate of τ is

$$\text{RSS}/(n - d), \tag{7.46}$$

the standard estimate for which the usual properties of maximum likelihood estimates apply provided $(n - d) \to \infty$.

It is hard to make totally precise the condition that L_1 contains little information about τ. Note that if β is a totally free parameter, in some sense, then an exact fit to any values of $v_1, \ldots, v_d$ can always be achieved in which case it is compelling that there is no base left for estimation of τ from this part of the original data. Any constraint, even a somewhat qualitative one, on β would, however, change the situation. Thus if it were thought that β is close to a fixed point, for example the origin, or more generally is near some space of lower dimension than d, and the distance of $\hat{\beta}$ from the origin or subspace is smaller than that to be expected on the basis of $\text{RSS}/(n-d)$, then there is some evidence that this latter estimate is too large.

We say that the likelihood $L_2(\tau)$ is *directly realizable* in that it is the ordinary likelihood, calculated under the original data-generating specification, of a system of observations that could have arisen. It is conceptually possible that the original data could have been collected, transformed and only $v_{d+1}, \ldots, v_n$ retained. In this and other similar cases, directly realizable likelihoods may be very different from the full likelihood of the whole data.

Such likelihoods are often obtained more generally by a preliminary transformation of Y to new variables V, W and then by writing the original likelihood in the form

$$f_V(v;\theta) f_{W|V}(w;v,\theta) \tag{7.47}$$

and arranging that one of the factors depends only on the parameter of interest ψ and that there are arguments for supposing that the information about ψ in the other factor is unimportant. Provided the transformation from Y to (V, W) does not depend on θ, the Jacobian involved in the transformation of continuous variables plays no role and will be ignored.

The most satisfactory version of this is when the factorization is *complete* in the sense that either

$$f_V(v;\lambda) f_{W|V}(w;v,\psi) \tag{7.48}$$

or

$$f_V(v;\psi) f_{W|V}(w;v,\lambda). \tag{7.49}$$

In the former case the information about ψ is captured by a *conditional likelihood* of W given V and in the latter case by the *marginal likelihood* of V.

There are other possible bases for factorizing the likelihood into components. When a factor can be obtained that totally captures the dependence on ψ there is no need for the component of likelihood to be directly realizable. The required first- and second-order properties of score and information follow directly from those of the whole log likelihood, as in the following example.

Example 7.11. *Parametric model for right-censored failure data.* For simplicity we consider a problem without explanatory variables. Let failure time have density $f_T(t;\psi)$ and survival function $S_T(t;\psi)$ and censoring time have density and survival function $f_C(c;\lambda)$ and $S_C(c;\lambda)$. We represent uninformative censoring by supposing that we have n independent pairs of independent random variables (T_k, C_k) and that we observe only $Y_k = \min(T_k, C_k)$ and the indicator variable $D_k = 1$ for a failure, i.e., $T_k < C_k$ and zero for censoring. Then the likelihood is

$$\Pi f_C^{1-d_k}(y_k; \lambda) S_C^{d_k}(y_k; \lambda) \times \Pi f_T^{d_k}(y_k; \psi) S_T^{1-d_k}(y_k; \psi). \tag{7.50}$$

Here we use the second factor for inference about ψ, even though it does not on its own correspond to a system of observations within the initial stochastic specification. Note that the specification assumes that there are no parameters common to the distributions of T and C. Were there such common parameters the analysis of the second factor would retain key properties, but in general lose some efficiency.

A factorization of the likelihood in which the values of the conditional information measure, $i_{\psi\psi\cdot\lambda}$, were respectively $O(1)$, i.e., bounded, and $O(n)$ would asymptotically have the same properties as complete separation.

7.6.5 Partial likelihood

We now discuss a particular form of factorization that is best set out in terms of a system developing in time although the temporal interpretation is not crucial. Suppose that, possibly after transformation, the sequence $Y_1, \ldots, Y_n$ of observed random variables can be written $T_1, S_1, \ldots, T_m, S_m$ in such a way that the density of S_k given C_k, all the previous random variables in the sequence, is specified as a function of ψ. Here m is related to but in general different from n. It is helpful to think of S_k as determining an event of interest and the T_k as determining associated relevant events that are, however, of no intrinsic interest for the current interpretation.

We then call

$$\Pi f_{S_k|C_k}(s_k \mid C_k = c_k; \psi) \tag{7.51}$$

a *partial likelihood* based on the S_k. Let U_k and $i_k(\psi)$ denote the contribution to the component score vector and information matrix arising from the kth term, i.e., from S_k. Then, because the density from which these are calculated is normalized to integrate to 1, we have that

$$E(U_k) = 0, \ \text{var}(U_k) = i_k(\psi). \tag{7.52}$$

Now the total partial likelihood is in general not normalized to 1 so that the same argument cannot be applied for the sum of the partial scores and information. We can, however, argue as follows. The properties of U_k hold conditionally on C_k, the total history up to the defining variable S_k, and hence

7.6 Modified likelihoods

in particular

$$E(U_k \mid U_1, \ldots, U_{k-1} = u_1, \ldots, u_{k-1}) = 0 \tag{7.53}$$

and, in particular for $l < k$,

$$E(U_l U_k) = E\{U_l E(U_k \mid U_l)\} = 0, \tag{7.54}$$

so that the components of the total score, while in general not independent, are uncorrelated. Therefore the total information matrix is the covariance matrix of the total score and, provided standard n asymptotics holds, the usual properties of maximum likelihood estimates and their associated statistics are achieved. In the more formal theory a martingale Central Limit Theorem will give the required asymptotic normality but, of course, the issue of the adequacy of the distributional approximation has always to be considered.

Example 7.12. *A fairly general stochastic process.* Consider a stochastic process with discrete states treated for simplicity in continuous time and let zero be a recurrent point. That is, on entry to state zero the system moves independently of its past and the joint density of the time to the next transition and the new state, k, occupied is $p_m(t; \psi)$ for $m = 1, 2, \ldots$. Recurrence means that it returns to zero state eventually having described a trajectory with properties of unspecified complexity. Then the product of the $p_m(t; \psi)$ over the exits from state zero, including, if necessary, a contribution from a censored exit time, obeys the conditions for a partial likelihood. If zero is a regeneration point, then each term in the partial likelihood is independent of the past and the partial likelihood is directly realizable, i.e., one could envisage an observational system in which only the transitions out of zero were observed. In general, however, each term $p_m(t; \psi)$ could depend in any specified way on the whole past of the process.

Example 7.13. *Semiparametric model for censored failure data.* We suppose that for each of n independent individuals observation of failure time is subject to uninformative censoring of the kind described in Example 7.11. We define, as before, the hazard function corresponding to a density of failure time $f(t)$ and survival function $S(t)$ to be $h(t) = f(t)/S(t)$. Suppose further that individual l has a hazard function for failure of

$$h_0(t) \exp(z_l^T \psi), \tag{7.55}$$

where z_l is a vector of possibly time-dependent explanatory variables for individual l, ψ is a vector of parameters of interest specifying the dependence of failure time on z and $h_0(t)$ is an unknown baseline hazard corresponding to an individual with $z = 0$.

Number the failures in order and let C_k specify all the information available about failures and censoring up to and including the time of the kth failure, including any changes in the explanatory variables, but excluding the identity of the individual who fails at this point. Let M_k specify the identity of that individual (simultaneous occurrences being assumed absent). Now C_k determines in particular $\mathcal{R}_k$, the risk set of individuals who are available to fail at the time point in question. Then the probability, given a failure at the time in question, that the failure is for individual M_k is

$$\exp\{z_{m_k}^T \psi\} h_0(t) / \Sigma_{l \in \mathcal{R}_k} \exp\{z_l^T \psi\} h_0(t). \tag{7.56}$$

Thus the baseline hazard cancels and the product of these expressions over k yields a relatively simple partial likelihood for ψ.

7.6.6 Pseudo-likelihood

We now consider a method for inference in dependent systems in which the dependence between observations is either only incompletely specified or specified only implicitly rather than explicitly, making specification of a complete likelihood function either very difficult or impossible. Many of the most important applications are to spatial processes but here we give more simple examples, where the dependence is specified one-dimensionally as in time series.

In any such fully specified model we can write the likelihood in the form

$$f_{Y_1}(y_1;\theta) f_{Y_2|Y_1}(y_2, y_1; \theta) \cdots f_{Y_n|Y^{(n-1)}}(y_n, y^{(n-1)}; \theta), \tag{7.57}$$

where in general $y^{(k)} = (y_1, \ldots, y_k)$. Now in a Markov process the dependence on $y^{(k)}$ is restricted to dependence on y_k and more generally for an m-dependent Markov process the dependence is restricted to the last m components.

Suppose, however, that we consider the function

$$\Pi f_{Y_k|Y_{k-1}}(y_k, y_{k-1}; \theta), \tag{7.58}$$

where the one-step conditional densities are correctly specified in terms of an interpretable parameter θ but where higher-order dependencies are not assumed absent but are ignored. Then we call such a function, that takes no account of certain dependencies, a *pseudo-likelihood*.

If $U_k(\theta)$ denotes the gradient or pseudo-score vector obtained from Y_k, because each term in the pseudo-likelihood is a probability density normalized to integrate to 1, then

$$E\{U_k(\theta); \theta\} = 0, \tag{7.59}$$

and the covariance matrix of a single vector U_k can be found from the appropriate information matrix. In general, however, the U_k are not independent and so the covariance matrix of the total pseudo-score is not $i(\theta)$, the sum of the separate information matrices. Provided that the dependence is sufficiently weak for standard n asymptotics to apply, the pseudo-maximum likelihood estimate $\tilde{\theta}$ found from (7.58) is asymptotically normal with mean θ and covariance matrix

$$i^{-1}(\theta)\text{cov}(U)i^{-1}(\theta). \tag{7.60}$$

Here $U = \Sigma U_k$ is the total pseudo-score and its covariance has to be found, for example by applying somewhat empirical time series methods to the sequence $\{U_k\}$.

Example 7.14. *Lag one correlation of a stationary Gaussian time series.* Suppose that the observed random variables are assumed to form a stationary Gaussian time series of unknown mean μ and variance σ^2 and lag one correlation ρ and otherwise to have unspecified correlation structure. Then Y_k given $Y_{k-1} = y_{k-1}$ has a normal distribution with mean $\mu + \rho(y_{k-1} - \mu)$ and variance $\sigma^2(1 - \rho^2)$; this last variance, the innovation variance in a first-order autoregressive process, may be taken as a new parameter. Note that if a term from Y_1 is included it will have the marginal distribution of the process, assuming observation starts in a state of statistical equilibrium. For inference about ρ the most relevant quantity is the adjusted score $U_{\rho \cdot \mu, \sigma}$ of (6.62) and its variance. Some calculation shows that the relevant quantity is the variance of

$$\Sigma\{(Y_k - \mu)(Y_{k-1} - \mu) - (Y_k - \mu)^2\}. \tag{7.61}$$

Here μ can be replaced by an estimate, for example the overall mean, and the variance can be obtained by treating the individual terms in the last expression as a stationary time series, and finding its empirical autocorrelation function.

Example 7.15. *A long binary sequence.* Suppose that $Y_1, \ldots, Y_m$ is a sequence of binary random variables all of which are mutually dependent. Suppose that the dependencies are represented by a latent multivariate normal distribution. More precisely, there is an unobserved random vector $W_1, \ldots, W_m$ having a multivariate normal distribution of zero means and unit variances and correlation matrix P and there are threshold levels $\alpha_1, \ldots, \alpha_m$ such that $Y_k = 1$ if and only if $W_k > \alpha_k$. The unknown parameters of interest are typically those determining P and possibly also the α. The data may consist of one long realization or of several independent replicate sequences.

The probability of any sequence of 0s and 1s, and hence the likelihood, can then be expressed in terms of the m-dimensional multivariate normal distribution function. Except for very small values of m there are two difficulties with this likelihood. One is computational in that high-dimensional integrals have to be evaluated. The other is that one might well regard the latent normal model as a plausible representation of low-order dependencies but be unwilling to base high-order dependencies on it.

This suggests consideration of the pseudo-likelihood

$$\Pi_{k>l} P(Y_k = y_k, Y_l = y_l), \tag{7.62}$$

which can be expressed in terms of the bivariate normal integral; algorithms for computing this are available.

The simplest special case is fully symmetric, having $\alpha_k = \alpha$ and all correlations equal, say to ρ. In the further special case $\alpha = 0$ of median dichotomy the probabilities can be found explicitly because, for example, by Sheppard's formula,

$$P(Y_k = Y_l = 0) = \frac{1}{4} + \frac{\sin^{-1}\rho}{2\pi}. \tag{7.63}$$

If we have n independent replicate sequences then first-, but not second-, order validity applies, as does standard n asymptotics, and estimates of α and ρ are obtained, the latter not fully efficient because of neglect of information in the higher-order relations.

Suppose, however, that m is large, and that there is just one very long sequence. It is important that in this case direct use of the pseudo-likelihood above is not satisfactory. This can be seen as follows. Under this special correlational structure we may write, at least for $\rho > 0$, the latent, i.e., unobserved, normal random variables, in the form

$$W_k = V\sqrt{\rho} + V_k\sqrt{(1-\rho)}, \tag{7.64}$$

where $V, V_1, \ldots, V_m$ are independent standard normal random variables. It follows that the model is equivalent to the use of thresholds at $\alpha - V\sqrt{\rho}$ with independent errors, instead of ones at α. It follows that if we are concerned with the behaviour for large m, the estimates of α and ρ will be close respectively to $\alpha - V\sqrt{\rho}$ and to zero. The general moral is that application of this kind of pseudo-likelihood to a single long sequence is unsafe if the long-range dependencies are not sufficiently weak.

Example 7.16. *Case-control study.* An important illustration of these ideas is provided by a type of investigation usually called a case-control study; in econometrics the term *choice-based sampling* is used. To study this it is helpful to consider as a preliminary a real or notional population of individuals, each having in principle a binary outcome variable y and two vectors of explanatory variables w and z. Individuals with $y = 1$ will be called *cases* and those with $y = 0$ *controls*. The variables w are to be regarded as describing the intrinsic

7.6 Modified likelihoods

nature of the individuals, whereas z are treatments or risk factors whose effect on y is to be studied. For example, w might include the gender of a study individual, z one or more variables giving the exposure of that individual to environmental hazards and y might specify whether or not the individual dies from a specific cause. Both z and w are typically vectors, components of which may be discrete or continuous.

Suppose that in the population we may think of Y, Z, W as random variables with

$$P(Y = 1 \mid W = w, Z = z) = L(\alpha + \beta^T z + \gamma^T w), \qquad (7.65)$$

where $L(x) = e^x/(1 + e^x)$ is the unit logistic function. Interest focuses on β, assessing the effect of z on Y for fixed w. It would be possible to include an interaction term between z and w without much change to the following discussion.

Now if the response $y = 1$ is rare and also if it is a long time before the response can be observed, direct observation of the system just described can be time-consuming and in a sense inefficient, ending with a very large number of controls relative to the number of cases. Suppose, therefore, instead that data are collected as follows. Each individual with outcome y, where $y = 0, 1$, is included in the data with conditional probability, given the corresponding z, w, of $p_y(w)$ and z determined retrospectively. For a given individual, $\mathcal{D}$ denotes its inclusion in the case-control sample. We write

$$P(\mathcal{D} \mid Y = y, Z = z, W = w) = p_y(w). \qquad (7.66)$$

It is crucial to the following discussion that the selection probabilities do not depend on z given w. In applications it would be quite common to take $p_1(w) = 1$, i.e., to take all possible cases. Then for each case one or more controls are selected with probability of selection defined by $p_0(w)$. Choice of $p_0(w)$ is discussed in more detail below.

It follows from the above specification that for the selected individuals, we have in a condensed notation that

$$P(Y = 1 \mid \mathcal{D}, z, w)$$
$$= \frac{P(Y = 1 \mid z, w) P(\mathcal{D} \mid 1, z, w)}{P(\mathcal{D} \mid z, w)}$$
$$= \frac{L(\alpha + \beta^T z + \gamma^T w) p_1(w)}{L(\alpha + \beta^T z + \gamma^T w) p_1(w) + \{1 - L(\alpha + \beta^T z + \gamma^T w)\} p_0(w)}$$
$$= L\{\alpha + \beta^T z + \gamma^T w + \log\{p_1(w)/p_0(w)\}\}. \qquad (7.67)$$

There are now two main ways of specifying the choice of controls. In the first, for each case one or more controls are chosen that closely match the case with respect to w. We then write for the kth such group of a case and controls, essentially without loss of generality,

$$\gamma^T w + \log\{p_1(w)/p_0(w)\} = \lambda_k, \tag{7.68}$$

where λ_k characterizes this group of individuals. This representation points towards the elimination of the nuisance parameters λ_k by conditioning. An alternative is to assume for the entire data that approximately

$$\log\{p_1(w)/p_0(w)\} = \eta + \zeta^T w \tag{7.69}$$

pointing to the unconditional logistic relation of the form

$$P(Y = 1 \mid \mathcal{D}, z, w) = L(\alpha^* + \beta^T z + \gamma^{*T} w). \tag{7.70}$$

The crucial point is that β, but not the other parameters, takes the same value as the originating value in the cohort study.

Now rather than base the analysis directly on one of the last two formulae, it is necessary to represent that in this method of investigation the variable y is fixed for each individual and the observed random variable is Z. The likelihood is therefore the product over the observed individuals of, again in a condensed notation,

$$f_{Z\mid\mathcal{D},Y,W} = f_{Y\mid\mathcal{D},Z,W} f_{Z\mid\mathcal{D},W} / f_{Y\mid\mathcal{D},W}. \tag{7.71}$$

The full likelihood function is thus a product of three factors. By (7.70) the first factor has a logistic form. The final factor depends only on the known functions $p_k(w)$ and on the numbers of cases and controls and can be ignored. The middle factor is a product of terms of the form

$$\frac{f_Z(z)\{L(\alpha + \beta^T z + \gamma^T w)p_1(w) + (1 - L(\alpha + \beta^T z + \gamma^T w))p_0(w)\}}{\int dv f_Z(v)\{L(\alpha + \beta^T v + \gamma^T w)p_1(w) + (1 - L(\alpha + \beta^T v + \gamma^T w))p_0(w)\}}. \tag{7.72}$$

Thus the parameter of interest, β, occurs both in the relatively easily handled logistic form of the first factor and in the second more complicated factor. One justification for ignoring the second factor could be to regard the order of the observations as randomized, without loss of generality, and then to regard the first logistic factor as a pseudo-likelihood. There is the difficulty, not affecting the first-order pseudo-likelihood, that the different terms are not independent in the implied new random system, because the total numbers of cases and controls are constrained, and may even be equal. Suppose, however, that the marginal distribution of Z, i.e., $f_Z(z)$, depends on unknown parameters ω in such a way that when $(\omega, \alpha, \beta, \gamma)$ are all unknown the second factor on its own provides no information about β. This would be the case if, for example, ω represented an arbitrary multinomial distribution over a discrete set of possible values of z. The profile likelihood for (α, β, γ), having maximized out over ω, is then essentially the logistic first factor which is thus the appropriate likelihood function for inference about β.

Another extreme case arises if $f_Z(z)$ is completely known, when there is in principle further information about β in the full likelihood. It is unknown whether such information is appreciable; it seems unlikely that it would be wise to use it.

7.6.7 Quasi-likelihood

The previous section concentrated largely on approximate versions of likelihood applicable when the dependency structure of the observations is relatively complicated. Quasi-likelihood deals with problems in which, while a relation between the mean and variance is reasonably secure, distributional assumptions are suspect.

The simplest instance is provided by Example 1.3, the linear model in which no normality assumption is made and an analysis is made on the basis of the second-order assumptions that the variance is constant and observations are mutually uncorrelated. The definition of the least squares estimates as obtained by orthogonal projection of Y on the space spanned by the columns of the defining matrix z and properties depending only on expectations of linear and quadratic functions of Y continue to hold.

The sufficiency argument used above to justify the optimality of least squares estimates does depend on normality but holds in weaker form as follows. In the linear model $E(Y) = z\beta$, any linear estimate of a single component of β must have the form $l^T Y$, where l is an *estimating vector*. Now l can be resolved into a component l_z in the column space of z and an orthogonal component $l_{\perp z}$. A direct calculation shows that the two components of $l^T Y$ are uncorrelated, so that $\text{var}(l^T Y)$ is the sum of two contributions. Also the second component has zero mean for all β. Hence for optimality we set $l_{\perp z} = 0$, the estimate has the form $l_z^T Y$, and, so long as z is of full rank, is uniquely the component of the least squares estimate. The proof may be called an appeal to *linear sufficiency*.

If instead of the components of Y being uncorrelated with equal variance, they have covariance matrix γV, where γ is a scalar, possibly unknown, and V is a known positive definite matrix, then the least squares estimates, defined by orthogonal projection in a new metric, become

$$z^T V^{-1}(Y - z\hat{\beta}) = 0, \qquad (7.73)$$

as can be seen, for example, by a preliminary linear transformation of Y to $V^{-1/2}Y$.

Now suppose that $E(Y) = \mu(\beta)$, where the vector of means is in general a nonlinear function of β, as well as of explanatory variables. We assume that the components of Y are uncorrelated and have covariance matrix again specified as $\gamma V(\beta)$, where γ is a constant and $V(\beta)$ is a known function of β. Now the analogue of the space spanned by the columns of z is the tangent space defined by

$$z^T(\beta) = \nabla \mu^T(\beta) \tag{7.74}$$

a $d_\beta \times n$ matrix; note that if $\mu(\beta) = z\beta$, then the new matrix reduces to the previous z^T. The estimating equation that corresponds to the least squares estimating equation is thus

$$z^T(\tilde\beta) V^{-1}(\tilde\beta)\{Y - \mu(\tilde\beta)\} = 0. \tag{7.75}$$

This may be called the *quasi-likelihood estimating equation*. It generalizes the linear least squares and the nonlinear normal-theory least squares equations.

The asymptotic properties of the resulting estimates can be studied via those of the *quasi-score function*

$$z^T(\beta) V^{-1}(\beta)\{Y - \mu(\beta)\}, \tag{7.76}$$

which mirror those of the ordinary score function for a parametric model. That score was defined as the gradient of the log likelihood and in many, but not all, situations the quasi-score is itself the gradient of a function $Q(\beta, y)$ which may be called the *quasi-likelihood*. Just as there are sometimes advantages to basing inference on the log likelihood rather than the score, so there may be advantages to using the quasi-likelihood rather than the quasi-score, most notably when the quasi-likelihood estimating equation has multiple roots.

By examining a local linearization of this equation, it can be shown that the properties of least squares estimates holding in the linear model under second-order assumptions extend to the quasi-likelihood estimates.

Relatively simple examples are provided by logistic regression for binary variables and log linear regression for Poisson variables. In the latter case, for example, the component observations are independent, the basic Poisson model has variances equal to the relevant mean, whereas the more general model has variances γ times the mean, thus allowing for overdispersion, or indeed for the less commonly encountered underdispersion.

Notes 7

Section 7.1. A saddle-point of a surface is a point such that some paths through it have local maxima and other paths local minima, as at the top of a mountain pass. The Hessian matrix at such a point has eigenvalues of both signs.

Section 7.2. A survey of nonregular problems is given by Smith (1989). Hall and Wang (2005) discuss Bayesian estimation of the end-point of a distribution.

The special choice of frequencies $\omega = \omega_p$ in Example 7.4 ensures such identities as

$$\Sigma \cos^2(\omega_p) = \Sigma \sin^2(\omega_p) = n/2,$$

as well as the exact orthogonality of components at different values of ω.

Section 7.3. The rather exceptional sets in fact have Lebesgue measure zero. The discussion here is based on Rotnitzky *et al.* (2000), where references to earlier work are given. The main application is to methods for dealing with so-called informative nonresponse. In some simpler situations with singular information a transformation of the parameter is all that is required.

Section 7.4. Cox (1961, 1962) develops a theory for estimates when the model is false in connection with the testing of separate families of hypotheses. This is concerned with testing a null hypothesis that the model lies in one family when the alternatives form a mathematically distinct although numerically similar family. The main applications have been econometric (White, 1994). See also Note 6.5.

Section 7.5. The argument for dealing with situations such as the normal mixture is due to Davies (1977, 1987). For a general discussion, see Ritz and Skovgaard (2005). In specific applications simulation is probably the best way to establish p-values.

Section 7.6. There is a very large literature on modified forms of likelihood, stemming in effect from Bartlett (1937) and then from Kalbfleisch and Sprott (1970). Partial likelihood was introduced by Cox (1972, 1975). The key property (7.54) links with the probabilistic notion of a martingale which underlies general discussions of maximum likelihood for dependent random variables (Silvey, 1961). Much work on survival analysis emphasizes this link (Andersen *et al.*, 1992).

Pseudo-likelihood was introduced by Besag (1974, 1977) and studied in a form related to the issues outlined here by Azzalini (1983). Example 7.5 is based on Cox and Reid (2004). Quasi-likelihood is due to Wedderburn (1974) and its optimal properties are given by McCullagh (1983). For a general treatment of modified likelihoods, see Lindsay (1988) and Varin and Vidoni (2005). Song *et al.* (2005) discussed breaking the log likelihood into two parts, one with relatively simple form and standard behaviour and the other encapsulating the complicated aspects. Formal application of these ideas in a Bayesian setting would allow specifications to be studied that were less complete than those needed for a full Bayesian discussion.

Other forms of likelihood centre more strongly on non- or semiparametric situations and include empirical likelihood (Owen, 2001) in which essentially multinomial distributions are considered with support the data values. There is a link with the nonparametric bootstrap. For a synthesis of some of these ideas, see Mykland (1995), who develops a notion of dual likelihood.

The fitting of smooth relations that are not specified parametrically is often accomplished by maximizing a formal likelihood function from which a correction is subtracted depending on the irregularity of the fitted relation. The correction might depend, for example, on an average squared value of a high-order derivative. While some arbitrariness may be attached both to the form of the penalty as well as to the weight attached to it, such penalized functions bring some unity to what would otherwise be a rather ad hoc issue. See Green and Silverman (1994).

The treatment of case-control studies largely follows Prentice and Pyke (1979) and Farewell (1979). For a comprehensive treatment of case-control studies, see Breslow and Day (1980). For a Bayesian analysis, see Seaman and Richardson (2004).

8

Additional objectives

Summary. This chapter deals in outline with a number of topics that fall outside the main theme of the book. The topics are prediction, decision analysis and point estimation, concentrating especially on estimates that are exactly or approximately unbiased. Finally some isolated remarks are made about methods, especially for relatively complicated models, that avoid direct use of the likelihood.

8.1 Prediction

In prediction problems the target of study is not a parameter but the value of an unobserved random variable. This includes, however, in so-called hierarchical models estimating the value of a random parameter attached to a particular portion of the data. In Bayesian theory the formal distinction between prediction and estimation largely disappears in that all unknowns have probability distributions. In frequentist theory the simplest approach is to use Bayes' theorem to find the distribution of the aspect of interest and to replace unknown parameters by good estimates. In special cases more refined treatment is possible.

In the special case when the value Y^*, say, to be predicted is conditionally independent of the data given the parameters the Bayesian solution is particularly simple. A predictive distribution is found by averaging the density $f_{Y^*}(y^*; \theta)$ over the posterior distribution of the parameter.

In special cases a formally exact frequentist predictive distribution is obtained by the following device. Suppose that the value to be predicted has the distribution $f_{Y^*}(y^*; \theta^*)$, whereas the data have the density $f_Y(y; \theta)$. Construct a sensible test of the null hypothesis $\theta = \theta^*$ and take all those values of y^* consistent with the null hypothesis at level c as the prediction interval or region.

Example 8.1. *A new observation from a normal distribution.* Suppose that the data correspond to independent and identically distributed observations having a normal distribution of unknown mean μ and known variance σ_0^2 and that it is required to predict a new observation from the same distribution. Suppose then that the new observation y^* has mean μ^*. Then the null hypothesis is tested by the standard normal statistic

$$\frac{y^* - \bar{y}}{\sigma_0 \sqrt{(1 + 1/n)}} \tag{8.1}$$

so that, for example, a level $1-c$ upper prediction limit for the new observation is

$$\bar{y} + k_c^* \sigma_0 \sqrt{(1 + 1/n)}. \tag{8.2}$$

The difference from the so-called plug-in predictor, $\bar{y} + k_c^* \sigma_0$, in which errors of estimating μ are ignored, is here slight but, especially if a relatively complicated model is used as the base for prediction, the plug-in estimate may seriously underestimate uncertainty.

8.2 Decision analysis

In many contexts data are analysed with a view to reaching a decision, for example in a laboratory science context about what experiment to do next. It is, of course, always important to keep the objective of an investigation in mind, but in most cases statistical analysis is used to guide discussion and interpretation rather than as an immediate automatic decision-taking device.

There are at least three approaches to a more fully decision-oriented discussion. Fully Bayesian decision theory supposes available a decision space $\mathcal{D}$, a utility function $U(d, \theta)$, a prior for θ, data y, and a model $f(y; \theta)$ for the data. The objective is to choose a decision rule maximizing for each y the expected utility averaged over the posterior distribution of θ. Such a rule is called a *Bayes rule*. The arguments for including a prior distribution are strong in that a decision rule should take account of all reasonably relevant information and not be confined to the question: what do these data plus model assumptions tell us? Here utility is defined in such a way that optimizing the expected utility is the appropriate criterion for choice. Utility is thus not necessarily expressed in terms of money and, even when it is, utility is not in general a linear function of monetary reward. In problems in which the consequences of decisions are long-lasting the further issue arises of how to discount future gains or losses of utility; exponential discounting may in some circumstances be inappropriate and some form of power-law discount function may be more suitable.

Wald's treatment of decision theory supposed that a utility function but not a prior is available; in some respects this is often an odd specification in that the choice of the utility may be at least as contentious as that of the prior. Wald showed that the only admissible decision rules, in a natural definition of admissibility, are Bayes rules and limits of Bayes rules. This leaves, of course, an enormous range of possibilities open; a minimax regret strategy is one possibility.

A third approach is to fall back on, for example, a more literal interpretation of the Neyman–Pearson theory of testing hypotheses and to control error rates at prechosen levels and to act differently according as y falls in different regions of the sample space. While there is substantial arbitrariness in the choice of critical rates in such an approach, it does achieve some consistency between different similar applications. In that sense it is sometimes used as an essential component by regulatory agencies, such as those dealing with new pharmaceutical products.

8.3 Point estimation

8.3.1 General remarks

The most obvious omission from the previous discussion is point estimation, i.e., estimation of a parameter of interest without an explicit statement of uncertainty. This involves the choice of one particular value when a range of possibilities are entirely consistent with the data. This is best regarded as a decision problem. With the exception outlined below imposition of constraints like *unbiasedness*, i.e., that the estimate T of ψ satisfy

$$E(T; \psi, \lambda) = \psi, \qquad (8.3)$$

for all (ψ, λ), is somewhat arbitrary.

Here we consider the point estimate on its own; this is distinct from taking it as the basis for forming a pivot, where some arbitrariness in defining the estimate can be absorbed into the distribution of the pivot.

An important generalization is that of an *unbiased estimating equation*. That is, the estimate is defined implicitly as the solution of the equation

$$g(Y, \tilde{\psi}) = 0, \qquad (8.4)$$

where for all θ we have

$$E\{g(Y, \psi); \theta\} = 0. \qquad (8.5)$$

Exact unbiasedness of an estimate, $E(T; \theta) = \psi$, is indeed rarely an important property of an estimate. Note, for example, that if T is exactly unbiased for ψ, then the estimate $h(T)$ will rarely be exactly unbiased for a nonlinear function $h(\psi)$. In any case our primary emphasis is on estimation by assessing an interval of uncertainty for the parameter of interest ψ. Even if this is conveniently specified via an implied pivot, a central value and a measure of uncertainty, the precise properties of the central value as such are not of direct concern.

There are, however, two somewhat related exceptions. One is where the point estimate is an intermediate stage in the analysis of relatively complex data. For example, the data may divide naturally into independent sections and it may be both convenient and enlightening to first analyse each section on its own, reducing it to one or more summarizing statistics. These may then be used as input into a model, quite possibly a linear model, to represent their systematic variation between sections, or may be pooled by averaging across sections. Even quite a small but systematic bias in the individual estimates might translate into a serious error in the final conclusion. The second possibility is that the conclusions from an analysis may need to be reported in a form such that future investigations could use the results in the way just sketched.

8.3.2 Cramér–Rao lower bound

A central result in the study of unbiased estimates, the Cramér–Rao inequality, shows the minimum variance achievable by an unbiased estimate. For simplicity suppose that θ is a scalar parameter and let $T = T(Y)$ be an estimate with bias $b(\theta)$ based on a vector random variable Y with density $f_Y(y; \theta)$. Then

$$\int t(y) f_Y(y; \theta) dy = \theta + b(\theta). \tag{8.6}$$

Under the usual regularity conditions, differentiate with respect to θ to give

$$\int t(y) U(y; \theta) f_Y(y; \theta) dy = 1 + b'(\theta), \tag{8.7}$$

where $U(y; \theta) = \partial \log f_Y(y; \theta) / \partial \theta$ is the gradient of the log likelihood. Because this gradient has zero expectation we can write the last equation

$$\text{cov}(T, U) = 1 + b'(\theta), \tag{8.8}$$

so that by the Cauchy–Schwarz inequality

$$\text{var}(T) \text{var}(U) \geq \{1 + b'(\theta)\}^2 \tag{8.9}$$

with equality attained if and only if T is a linear function of U; see Note 8.2. Now $\text{var}(U)$ is the expected information $i(\theta)$ in the data y. Thus, in the particular case of unbiased estimates,

$$\text{var}(T) \geq 1/i(\theta). \tag{8.10}$$

A corresponding argument in the multiparameter case gives for the variance of an unbiased estimate of, say, the first component of θ the lower bound $i^{11}(\theta)$, the (1, 1) element of the inverse of the expected information matrix.

The results here concern the variance of an estimate. The role of information in the previous chapter concerned the asymptotic variance of estimates, i.e., the variance of the limiting normal distribution, not in general to be confused with the limiting form of the variance of the estimate.

Example 8.2. *Exponential family.* In the one-parameter exponential family with density $m(y) \exp\{s\phi - k(\phi)\}$ the log likelihood derivative in the canonical parameter ϕ is linear in s and hence any linear function of S is an unbiased estimate of its expectation achieving the lower bound. In particular this applies to S itself as an estimate of the moment parameter $\eta = k'(\theta)$.

Thus if $Y_1, \ldots, Y_n$ are independently normally distributed with mean μ and known variance σ_0^2 the mean $\bar{Y}$ is unbiased for μ and its variance achieves the lower bound σ_0^2/n. If, however, the parameter of interest is $\theta = \mu^2$, the estimate $\bar{Y}^2 - \sigma_0^2/n$ is unbiased but its variance does not achieve the lower bound. More seriously, the estimate is not universally sensible in that it estimates an essentially nonnegative parameter by a quantity that may be negative, in fact with appreciable probability if $|\mu|$ is very small. The above estimate could be replaced by zero in such cases at the cost of unbiasedness. Note, however, that would be unwise if the estimate is to be combined linearly with other estimates from independent data.

Example 8.3. *Correlation between different estimates.* Suppose that T and T^* are both exactly (or very nearly) unbiased estimates of the same parameter θ and that T has minimum variance among unbiased estimates. Then for any constant a the estimate $aT + (1-a)T^*$ is also unbiased and has variance

$$a^2 \text{var}(T) + 2a(1-a)\text{cov}(T, T^*) + (1-a)^2 \text{var}(T^*). \tag{8.11}$$

This has its minimum with respect to a at $a = 1$ implying that

$$\text{cov}(T, T^*) = \text{var}(T), \tag{8.12}$$

as in a sense is obvious, because the estimate T^* may be regarded as T plus an independent estimation error of zero mean.

Alternatively

$$\text{corr}(T, T^*) = \{\text{var}(T)/\text{var}(T^*)\}^{1/2}. \tag{8.13}$$

The ratio of variances on the right-hand side may reasonably be called the *efficiency of T^* relative to T*, so that the correlation of an inefficient estimate with a corresponding minimum variance estimate is the square root of the efficiency. For example, it may be shown that the median of a set of n independently normally distributed random variables of mean μ and variance σ_0^2 is an unbiased estimate of μ with variance approximately $\pi\sigma_0^2/(2n)$. That is, the correlation between mean and median is $\sqrt{2}/\sqrt{\pi}$, which is approximately 0.8.

Essentially the same idea can be used to compare the efficiency of tests of significance and to study the correlation between two different test statistics of the same null hypothesis. For this suppose that T and T^*, instead of being estimates of θ, are test statistics for a null hypothesis $\theta = \theta_0$ and such that they are (approximately) normally distributed with means $\mu(\theta)$ and $\mu^*(\theta)$ and variances $\sigma_T^2(\theta)/n$ and $\sigma_{T^*}^2(\theta)/n$. Now locally near the hypothesis we may linearize the expectations writing, for example

$$\mu(\theta) = \mu(\theta_0) + [d\mu(\theta)/d\theta]_{\theta_0}(\theta - \theta_0), \tag{8.14}$$

thereby converting the test statistic into a local estimate

$$\{T - \mu(\theta_0)\}\{[d\mu(\theta)/d\theta]_{\theta_0}\}^{-1}. \tag{8.15}$$

Near the null hypothesis this estimate has approximate variance

$$\{\sigma^2(\theta_0)/n\}\{[d\mu(\theta)/d\theta]_{\theta_0}\}^{-2}. \tag{8.16}$$

This leads to the definition of the *efficacy* of the test as

$$\{[d\mu(\theta)/d\theta]_{\theta_0}\}^2\{\sigma^2(\theta_0)\}^{-1}. \tag{8.17}$$

The *asymptotic efficiency* of T^* relative to the most efficient test T is then defined as the ratio of the efficacies. Furthermore, the result about the correlation between two estimates applies approximately to the two test statistics. The approximations involved in this calculation can be formalized via considerations of asymptotic theory.

Example 8.4. *The sign test.* Suppose that starting from a set of independent normally distributed observations of unknown mean θ and known variance σ_0^2 we consider testing the null hypothesis $\theta = \theta_0$ and wish to compare the standard test based on the mean $\bar{Y}$ with the sign test, counting the number of observations greater than θ_0. The former leads to the test statistic

$$T = \bar{Y} \tag{8.18}$$

with $d\mu(\theta)/d\theta = 1$ and efficacy $1/\sigma_0^2$. The sign test uses T^* to be the proportion of observations above θ_0 with binomial variance $1/(4n)$ at the null hypothesis and with

$$\mu^*(\theta) = \Phi(\theta - \theta_0) - 1/2, \qquad (8.19)$$

so that the efficacy is $2/(\pi\sigma^2)$. Thus the asymptotic efficiency of the sign test relative to the test based on the mean is $2/\pi$. Not surprisingly, this is the ratio of the asymptotic variances of the mean and median as estimates of μ, as is indeed apparent from the way efficacy was defined.

8.3.3 Construction of unbiased estimates

Occasionally it is required to find a function of a sufficient statistic, S, that is, exactly or very nearly unbiased for a target parameter θ. Let V be any unbiased estimate of θ and let S be a sufficient statistic for the estimation of θ; we appeal to the property of completeness mentioned in Section 4.4. Then $T = E(V \mid S)$ is a function of S unbiased for θ and having minimum variance. Note that because the distribution of Y given S does not involve θ neither does T.

The mean of T is

$$E_S E(V \mid S) = E(V) = \theta. \qquad (8.20)$$

To find $\text{var}(T)$ we argue that

$$\text{var}(V) = \text{var}_S E(V \mid S) + \text{var}_S E(V \mid S). \qquad (8.21)$$

The first term on the right-hand side is $\text{var}(T)$ and the second term is zero if and only if V is a function of S. Thus T has minimum variance. Completeness of S in the sense that no function of S can have zero expectation for all θ shows that T is unique, i.e., two different V must lead to the same T.

A second possibility is that we have an approximately unbiased estimate and wish to reduce its bias further. The easiest way to do this appeals to asymptotic considerations and properly belongs to Chapter 6, but will be dealt with briefly here. Commonly the initial estimate will have the form $h(T)$, where $h(.)$ is a given function, $h(\theta)$ is the target parameter and T has reasonably well-understood properties, in particular its expectation is θ and its variance is known to an adequate approximation.

Example 8.5. *Unbiased estimate of standard deviation.* Suppose that $Y_1, \ldots, Y_n$ are independently distributed with the same mean μ, the same variance $\tau = \sigma^2$ and the same fourth cumulant ratio $\gamma_4 = \kappa_4/\tau^2 = \mu_4/\tau^2 - 3$, where κ_4 and μ_4 are respectively the fourth cumulant and the fourth moment about the mean of the distribution of the Y_k; note that the notation γ_4 is preferred to the more commonly used notation γ_2 or the older $\beta_2 - 3$. If s^2 is the standard estimate of variance, namely $\Sigma(Y_k - \bar{Y})^2/(n-1)$, direct calculation with expectations shows that s^2 is exactly unbiased for τ and that approximately for large n

$$\text{var}(s^2) = 2(1 + \gamma_4/2)\tau^2/n. \tag{8.22}$$

To obtain an approximately unbiased estimate of σ, write $T = s^2$ and consider $\sqrt{T}$. We expand about τ to obtain

$$\sqrt{T} = \sqrt{\tau}\{1 + (T-\tau)/\tau\}^{1/2}$$
$$= \sqrt{\tau}\{1 + (T-\tau)/(2\tau) - (T-\tau)^2/(8\tau^2) + \cdots\}. \tag{8.23}$$

Now take expectations, neglecting higher-order terms. There results

$$E(\sqrt{T}) = \sqrt{\tau}\{1 - (1 + \gamma_4/2)/(8n)\}, \tag{8.24}$$

suggesting, recalling that $s = \sqrt{T}$, the revised estimate

$$s\{1 + (1 + \gamma_4/2)/(8n)\}. \tag{8.25}$$

This requires either some approximate knowledge of the fourth cumulant ratio γ_4 or its estimation, although the latter requires a large amount of data in the light of the instability of estimates of fourth cumulants.

For a normal originating distribution the correction factor is to this order $\{1 + 1/(8n)\}$. Because in the normal case the distribution of T is proportional to chi-squared with $n-1$ degrees of freedom, it follows that

$$E(\sqrt{T}) = \sqrt{\tau}\frac{\sqrt{2}\Gamma(n/2)}{\sqrt{(n-1)}\Gamma(n/2 - 1/2)} \tag{8.26}$$

from which an exactly unbiased estimate can be constructed. The previous argument essentially, in this case, amounts to a Laplace expansion of the ratio of gamma functions.

This argument has been given in some detail because it illustrates a simple general method; if the originating estimate is a function of several component random variables a multivariable expansion will be needed. It is to be stressed that the provision of exactly (or very nearly) unbiased estimates is rarely, if ever, important in its own right, but it may be part of a pathway to more important goals.

As a footnote to the last example, the discussion raises the relevance of the stress laid in elementary discussions of estimation on the importance of dividing a sum of squares of deviations from the mean by $n-1$ and not by n in order to estimate variance. There are two justifications. One is that it eases the merging of estimates of variance by pooling small samples from different sources; the pooled sums of squares are divided by the pooled degrees of freedom. The other reason is more one of convention. By always dividing a residual sum of squares from a linear model by the residual degrees of freedom, i.e., by the rank of the relevant quadratic form, comparability with, for example, the use of the Student t distribution for inference about a regression coefficient is achieved. The argument that an individual estimate is unbiased has little force in that the sampling error is much greater than any bias induced by dividing by, say, n rather than by $n-1$. In general for a single estimate adjustments that are small compared with the sampling error are unimportant; when several or many such estimates are involved the position is different.

Note finally that while comparison of asymptotic variances of alternative estimates is essentially a comparison of confidence intervals, and is invariant under nonlinear transformation of the target parameter, such stability of interpretation does not hold for small-sample comparisons either of mean squared error or of variances of bias-adjusted estimates.

8.4 Non-likelihood-based methods

8.4.1 General remarks

A limitation, at least superficially, on the discussion in Chapters 1–7 has been its strong emphasis on procedures derived from likelihood and gaining both intuitive appeal and formal strength thereby. While, especially with modern computational developments, tackling new problems via some form of likelihood is often very attractive, nevertheless there are many contexts in which some less formal approach is sensible. This and the following section therefore consider methods in which likelihood is not used, at least not explicitly.

Likelihood plays two roles in the earlier discussion. One is to indicate which aspects of the data are relevant and the other, especially in the Bayesian approach, is to show directly how to extract interpretable inference. In some of the situations to be outlined, it may be that some other considerations indicate summarizing features of the data. Once that step has been taken likelihood-based ideas may be used as if those summarizing features were the only data available. A possible loss of efficiency can sometimes then be assessed by, for example, comparing the information matrix from the reduced data with that from the originating data. Most of the previous discussion is then available.

Some of the reasons for not using the likelihood of the full data are as follows:

- obtaining the likelihood in useful form may be either impossible or prohibitively time-consuming;
- it may be desired to express a dependency via a smooth curve or surface not of prespecified parametric form but obtained via some kind of smoothing method;
- it may be desired to make precision assessment and significance testing by cautious methods making no parametric assumption about the form of probability distributions of error;
- there may be a need to use procedures relatively insensitive to particular kinds of contamination of the data;
- it may be wise to replace complex computations with some more transparent procedure, at least as a preliminary or for reassurance;
- there may be a need for very simple procedures, for example for computational reasons, although computational savings as such are less commonly important than was the case even a few years ago.

8.4.2 Alternatives to the log likelihood

We begin with a relatively minor modification in which the log likelihood $l(\theta)$ is replaced by a function intended to be essentially equivalent to it but in a sense simpler to handle. To the order of asymptotic theory considered here $l(\theta)$ could be replaced by any function having its maximum at $\hat{\theta} + O_p(1/n)$ and having second derivatives at the maximum that are $\hat{j}\{1 + O_p(1/\sqrt{n})\}$. There are many such functions leading to estimates, confidence limits and tests that are all in a sense approximately equivalent to those based on $l(\theta)$. That is, the estimates typically differ by $O_p(1/n)$, a small difference when compared with the standard errors which are $O_p(1/\sqrt{n})$.

In fact a number of considerations, notably behaviour in mildly anomalous cases and especially properties derived from higher-order expansions, suggest that log likelihood is the preferred criterion. Sometimes, however, it is useful to consider other possibilities and we now consider one such, reduction to the method of least squares.

For this we suppose that we have m random variables $Y_1, \ldots, Y_m$ and that each is a condensation of more detailed data, each Y_k having a large information content, i.e., $O(n)$. A fuller notation would replace Y_k, say, by $Y_k^{(n)}$ to emphasize that the asymptotics we consider involve not increasing m but rather increasing internal replication within each Y_k.

8.4 Non-likelihood-based methods

Suppose that there is a transformation from Y_k to $h(Y_k)$ such that asymptotically in n the transformed variable is normally distributed with mean a simple function of θ and variance τ_k which can be estimated by v_k such that $v_k/\tau_k = 1 + O_p(1/\sqrt{n})$. Although the restriction is not essential, we shall take the simple function of θ to be linear, and using the notation of the linear model, write the $m \times 1$ vector of asymptotic means in the form

$$E\{h(Y)\} = z\beta, \qquad (8.27)$$

where z is an $m \times p$ matrix of constants of full rank $p < m$ and β is a $p \times 1$ vector of parameters and cov$\{h(Y)\}$ is consistently estimated by diag(v_k). This is to be taken as meaning that the log likelihood from the formal linear model just specified agrees with that from whatever model was originally specified for Y with error $O_p(1/\sqrt{n})$. Note for this that the log likelihood for $h(Y_k)$ differs by at most a constant from that for Y_k and that typically a random variable whose asymptotic distribution is standard normal has a distribution differing from the standard normal by $O(1/\sqrt{n})$. In this argument, as noted above, the asymptotics, and hence the approximations, derive from internal replication within each Y_k, not from notional increases in m.

Application of standard weighted least squares methods to (8.27) may be called *locally linearized least squares with empirically estimated weights*.

Example 8.6. *Summarization of binary risk comparisons.* Suppose that the primary data are binary outcomes for individuals who have one of two possible levels of a risk factor or treatment. In the kth investigation for $k = 1, \ldots, m$, there are n_{kt} individuals in risk category t for $t = 0, 1$ of whom r_{kt} show a positive response. If π_{kt} denotes the corresponding probability of a positive response, we have previously in Example 4.5 considered the linear logistic model in which

$$\log\{\pi_{kt}/(1-\pi_{kt})\} = \alpha_{kt} + \Delta t. \qquad (8.28)$$

There are arbitrary differences between levels of k but the logistic difference between $t = 0$ and $t = 1$ is constant.

We take Y_k to be the collection $(n_{k0}, R_{k0}, n_{k1}, R_{k1})$, where the R_{kt} are binomial random variables. Let

$$h(Y_k) = \log \frac{R_{k1}}{R_{k0}} \frac{(n_{k0} - R_{k0})}{(n_{k1} - R_{k1})}. \qquad (8.29)$$

This may be called the *empirical logistic difference* between the two conditions because it is a difference of two observed log odds, each of which is an empirical logistic transform.

Then asymptotically $h(Y_k)$ is normally distributed with mean Δ and variance estimated by

$$v_k = 1/R_{k1} + 1/R_{k0} + 1/(n_{k1} - R_{k1}) + 1/(n_{k0} - R_{k0}). \tag{8.30}$$

Weighted least squares gives for the estimation of the single parameter Δ the estimate and asymptotic variance

$$\tilde{\Delta} = \frac{\Sigma h(Y_k) v_k^{-1}}{\Sigma v_k^{-1}} \tag{8.31}$$

and

$$1/\Sigma v_k^{-1}. \tag{8.32}$$

Moreover, the weighted residual sum of squares, namely

$$\Sigma \{h(Y_k) - \tilde{\Delta}\}^2 / v_k, \tag{8.33}$$

has an asymptotic chi-squared distribution with $m - 1$ degrees of freedom and is the asymptotic equivalent to the likelihood ratio and other tests of the constancy of the logistic difference for all k.

The following further points arise.

- Other functions than the logistic difference could be used, for example the estimated difference between probabilities themselves.
- If significant heterogeneity is found via the chi-squared test, the next step is to find relevant explanatory variables characterizing the different k and to insert them as explanatory variables in the linear model.
- If it is suspected that the binomial variation underlying the calculation of v_k underestimates the variability, inflation factors can be applied directly to the v_k.
- If the data emerging for each k and encapsulated in Y_k include adjustments for additional explanatory variables, the argument is unchanged, except possibly for minor changes to v_k to allow for the additional variance attached to the adjustments.
- The standard subsidiary methods of analysis associated with the method of least squares, for example those based on residuals, are available.

Any advantage to this method as compared with direct use of the log likelihood probably lies more in its transparency than in any computational gain from the avoidance of iterative calculations.

8.4.3 Complex models

One way of fitting a complex model, for example a stochastic model of temporal or spatial-temporal development, is to find which properties of the model can be evaluated analytically in reasonably tractable form. For a model with a d-dimensional parameter, the simplest procedure is then to equate d functions of the data to their analytical analogues and to solve the resulting nonlinear equations numerically. If there is a choice of criteria, ones judged on some intuitive basis as likely to be distributed independently are chosen. For example, from data on daily rainfall one might fit a four-parameter stochastic model by equating observed and theoretical values of the mean and variance, the proportion of dry days and the lag one serial correlation coefficient.

If s further features can be evaluated analytically they can be used for an informal test of adequacy, parallelling informally the sufficiency decomposition of the data used in simpler problems. Alternatively, the $d + s$ features could be fitted by the d parameters using some form of weighted nonlinear generalized least squares. For this a covariance matrix would have to be assumed or estimated by simulation after a preliminary fit.

Such methods lead to what may be called *generalized method of moments estimates*, named after the parallel with a historically important method of fitting parametric frequency curves to data in which for d parameters the first d moments are equated to their theoretical values.

Especially if the features used to fit the model are important aspects of the system, such methods, although seemingly rather empirical compared with those developed earlier in the book, may have appreciable appeal in that even if the model is seriously defective at least the fitted model reproduces important properties and might, for example, therefore be a suitable base for simulation of further properties.

An important feature in the fitting of relatively complex models, especially stochastic models of temporal or spatial-temporal data, is that it may be recognized that certain features of the data are poorly represented by the model, but this may be considered largely irrelevant, at least provisionally. Such features might dominate a formal likelihood analysis of the full data, which would therefore be inappropriate.

Such issues are particularly potent for stochastic models developed in continuous time. For formal mathematical purposes, although not for rigorous general development, continuous time processes are often easier to deal with than the corresponding discrete time models. Whenever a continuous time model is fitted to empirical data it is usually important to consider critically the range of time scales over which the behaviour of the model is likely to be a reasonable approximation to the real system.

For example, the sample paths of the stochastic process might contain piece-wise deterministic sections which might allow exact determination of certain parameters if taken literally. Another somewhat similar issue that may arise especially with processes in continuous time is that the very short time scale behaviour of the model may be unrealistic, without prejudicing properties over wider time spans.

Example 8.7. *Brownian motion.* Suppose that the model is that the process $\{Y(t)\}$ is Brownian motion with zero drift and unknown variance parameter (diffusion coefficient) σ^2. That is, increments of the process over disjoint intervals are assumed independently normally distributed with for any $(t, t+h)$ the increment $Y(t+h) - Y(t)$ having zero mean and variance $\sigma^2 h$. If we observe $Y(t)$ over the interval $(0, t_0)$, we may divide the interval into t_0/h nonoverlapping subintervals each of length h and from each obtain an unbiased estimate of σ^2 of the form

$$\{Y(kh+h) - Y(kh)\}^2/h. \tag{8.34}$$

There being t_0/h independent such estimates, we obtain on averaging an estimate with t_0/h degrees of freedom. On letting $h \to 0$ it follows that for any t_0, no matter how small, σ^2 can be estimated with arbitrarily small estimation error. Even though Brownian motion may be an excellent model for many purposes, the conclusion is unrealistic, either because the model does not hold over very small time intervals or because the measurement process distorts local behaviour.

8.4.4 Use of notionally inefficient statistics

As noted in the preliminary remarks, there are various reasons for using other than formally efficient procedures under a tightly specified model. In particular there is an extensive literature on robust estimation, almost all concentrating on controlling the influence of observations extreme in some sense. Derivation of such estimates can be done on an intuitive basis or by specifying a new model typically with modified tail behaviour. Estimates can in many cases be compared via their asymptotic variances; a crucial issue in these discussions is the preservation of the subject-matter interpretation of the parameter of interest.

More generally, frequentist-type analyses lend themselves to the theoretical comparison of alternative methods of analysis and to the study of the sensitivity of methods to misspecification errors. This may be done by mathematical analysis or by computer simulation or often by a combination. It is common to present the conclusions of such studies by one or more of:

- the bias, variance and possibly mean square error of alternative point estimates;
- the coverage properties, at a given level, of confidence intervals;
- the probability of rejecting correct null hypotheses, and of power, i.e., the probability of detecting specified departures, both types of comparison being at prespecified levels, for example 5 per cent or 1 per cent values of p.

These are, however, quite often not the most revealing summaries of, for example, extensive simulation studies. For instance, coverage properties may better be studied via the whole distributions of pivots based on alternative analyses and significance tests similarly by comparing distributions of the random variable P.

In terms of the analysis of particular sets of data the issue is whether specific conclusions are vulnerable to underlying assumptions and sensitivity assessment has to be appropriately conditional.

Parallel Bayesian analyses involve the sensitivity of a posterior distribution to perturbation of either the likelihood or the prior and again can be studied analytically or purely numerically.

Notes 8

Section 8.1. Some accounts of statistical analysis put much more emphasis than has been done in this book on prediction as contrasted with parameter estimation. See, for example, Geisser (1993) and Aitchison and Dunsmore (1975). Many applications have an element of or ultimate objective of prediction. Often, although certainly not always, this is best preceded by an attempt at analysis and understanding. For a discussion of prediction from a frequentist viewpoint, see Lawless and Fridette (2005). An alternative approach is to develop a notion of predictive likelihood. For one such discussion, see Butler (1986).

A series of papers summarized and substantially developed by Lee and Nelder (2006) have defined a notion of h-likelihood in the context of models with several layers of error and have applied it to give a unified treatment of a very general class of models with random elements in both systematic and variance structure. Some controversy surrounds the properties that maximizing the h-likelihood confers.

Section 8.2. An element of decision-making was certainly present in the early work of Neyman and Pearson, a Bayesian view being explicit in Neyman and

Pearson (1933), and the decision-making aspect of their work was extended and encapsulated in the attempt of Wald (1950) to formulate all statistical problems as ones of decision-making. He assumed the availability of a utility (or loss) function but no prior probability distribution, so that Bayesian decision rules played a largely technical role in his analysis. Most current treatments of decision theory are fully Bayesian, in principle at least. There is a very strong connection with economic theory which does, however, tend to assume, possibly after a largely ritual hesitation, that utility is linear in money.

Section 8.3. Emphasis has not been placed on point estimation, especially perhaps on point estimates designed to minimize mean squared error. The reason is partly that squared error is often not a plausible measure of loss, especially in asymmetric situations, but more importantly the specification of isolated point estimates without any indication of their precision is essentially a decision problem and should be treated as such. In most applications in which point estimates are reported, either they are an intermediate step in analysis or they are essentially the basis of a pivot for inference. In the latter case there is some allowable arbitrariness in the specification.

The Cauchy–Schwarz inequality in the present context is proved by noting that, except in degenerate cases, for all constants a, b the random variable $aT + bU$ has strictly positive variance. The quadratic equation for, say, b/a obtained by equating the variance to zero thus cannot have a real solution.

The elegant procedure for constructing unbiased estimates via an arbitrary unbiased estimate and a sufficient statistic is often called Rao–Blackwellization, after its originators C. R. Rao and D. Blackwell. A collection of papers on estimating equations was edited by Godambe (1991) and includes a review of his own contributions. An especially influential contribution is that of Liang and Zeger (1986).

Example 8.3 follows an early discussion of Fisher (1925a). The notion of asymptotic relative efficiency and its relation to that of tests was introduced essentially in the form given here by Cochran (1937), although it was set out in more formal detail in lecture notes by E. J. G. Pitman, after whom it is usually named. There are other ideas of relative efficiency related to more extreme tail behaviour.

Section 8.4. Estimates with the same asymptotic distribution as the maximum likelihood estimate to the first order of asymptotic theory are called BAN (best asymptotically normal) or if it is desired to stress some regularity conditions RBAN (regular best asymptotically normal). The idea of forming summarizing statistics on the basis of some mixture of judgement and ease of interpretation and then using these as the base for a second more formal stage of analysis has a long history, in particular in the context of longitudinal data. In econometrics it has been developed as the generalized method of moments. For an account with examples under the name 'the indirect method', see Jiang and Turnbull (2004). The asymptotic variance stated for the log odds is obtained by using local linearization (the so-called delta method) to show that for any function $h(R/n)$ differentiable at μ, then asymptotically $\text{var}\{h(R/n)\} = \{h'(\mu)\}^2 \text{var}(R/n)$, where $\mu = E(R/n)$.

The more nearly linear the function $h(.)$ near μ the better the approximation.

9
Randomization-based analysis

Summary. A different approach to statistical inference is outlined based not on a probabilistic model of the data-generating process but on the randomization used in study design. The implications of this are developed in simple cases, first for sampling and then for the design of experiments.

9.1 General remarks

The discussion throughout the book so far rests centrally on the notion of a probability model for the data under analysis. Such a model represents, often in considerably idealized form, the data-generating process. The parameters of interest are intended to capture important and interpretable features of that generating process, separated from the accidental features of the particular data. That is, the probability model is a model of physically generated variability, of course using the word 'physical' in some broad sense. This whole approach may be called *model-based*.

In some contexts of sampling existing populations and of experimental design there is a different approach in which the probability calculations are based on the randomization used by the investigator in the planning phases of the investigation. We call this a *design-based* formulation.

Fortunately there is a close similarity between the methods of analysis emerging from the two approaches. The more important differences between them concern interpretation of the conclusions. Despite the close similarities it seems not to be possible to merge a theory of the purely design-based approach seamlessly into the theory developed earlier in the book. This is essentially because of the absence of a viable notion of a likelihood for the haphazard component of the variability within a purely design-based development.

9.2 Sampling a finite population

These points will be illustrated in terms of two very simple examples, one of sampling and one of experimental design.

9.2 Sampling a finite population

9.2.1 First and second moment theory

Suppose that we have a well-defined population of N labelled individuals and that the sth member of the population has a real-valued property η_s. The quantity of interest is assumed to be the finite population mean $m_\eta = \Sigma \eta_s / N$. The initial objective is the estimation of m_η together with some assessment of the precision of the resulting estimate. More analytical and comparative aspects will not be considered here.

To estimate m_η suppose that a sample of n individuals is chosen at random without replacement from the population on the basis of the labels. That is, some impersonal sampling device is used that ensures that all $N!/\{n!(N-n)!\}$ distinct possible samples have equal probability of selection. Such *simple random sampling without replacement* would rarely be used directly in applications, but is the basis of many more elaborate and realistic procedures. For convenience of exposition we suppose that the order of the observations $y_1, \ldots, y_n$ is also randomized.

Define an indicator variable I_{ks} to be 1 if the kth member of the sample is the sth member of the population. Then in virtue of the sampling method, not in virtue of an assumption about the structure of the population, the distribution of the I_{ks} is such that

$$E_R(I_{ks}) = P_R(I_{ks} = 1) = 1/N, \tag{9.1}$$

and, because the sampling is without replacement,

$$E_R(I_{ks}I_{lt}) = (1 - \delta_{kl})(1 - \delta_{st})/\{N(N-1)\} + \delta_{kl}\delta_{st}/N, \tag{9.2}$$

where δ_{kl} is the Kronecker delta symbol equal to 1 if $k = l$ and 0 otherwise. The suffix R is to stress that the probability measure is derived from the sampling randomization. More complicated methods of sampling would be specified by the properties of these indicator random variables.

It is now possible, by direct if sometimes tedious calculation from these and similar higher-order specifications of the distribution, to derive the moments of the sample mean and indeed any polynomial function of the sample values. For this we note that, for example, the sample mean can be written

$$\bar{y} = \Sigma_{k,s} I_{ks} \eta_s / n. \tag{9.3}$$

It follows immediately from (9.1) that

$$E_R(\bar{y}) = m_\eta, \quad (9.4)$$

so that the sample mean is unbiased in its randomization distribution. Similarly

$$\mathrm{var}_R(\bar{y}) = \Sigma_{k,s}\mathrm{var}(I_{k,s})\eta^2/n^2 + 2\Sigma_{k>l,s>t}\mathrm{cov}(I_{ks}, I_{lt})\eta_s\eta_t/n^2. \quad (9.5)$$

It follows from (9.2) that

$$\mathrm{var}_R(\bar{y}) = (1-f)v_{\eta\eta}/n, \quad (9.6)$$

where the second-moment variability of the finite population is represented by

$$v_{\eta\eta} = \Sigma(\eta_s - m_\eta)^2/(N-1), \quad (9.7)$$

sometimes inadvisably called the finite population variance, and $f = n/N$ is the sampling fraction.

Thus we have a simple generalization of the formula for the variance of a sample mean of independent and identically distributed random variables. Quite often the proportion f of individuals sampled is small and then the factor $1-f$, called the *finite population correction*, can be omitted.

A similar argument shows that if $s^2 = \Sigma(y_k - \bar{y})^2/(n-1)$, then

$$E_R(s^2) = v_{\eta\eta}, \quad (9.8)$$

so that the pivot

$$\frac{m_\eta - \bar{y}}{\{s^2(1-f)/n\}^{1/2}} \quad (9.9)$$

has the form of a random variable of zero mean divided by an estimate of its standard deviation. A version of the Central Limit Theorem is available for this situation, so that asymptotically confidence limits are available for m_η by pivotal inversion.

A special case where the discussion can be taken further is when the population values η_s are binary, say 0 or 1. Then the number of sampled individuals having value 1 has a hypergeometric distribution and the target population value is the number of 1s in the population, a defining parameter of the hypergeometric distribution and in principle in this special case design-based inference is equivalent, formally at least, to parametric inference.

9.2.2 Discussion

Design-based analysis leads to conclusions about the finite population mean totally free of assumptions about the structure of the variation in the population

9.2 Sampling a finite population

and subject, for interval estimation, only to the usually mild approximation of normality for the distribution of the pivot. Of course, in practical sampling problems there are many complications; we have ignored the possibility that supplementary information about the population might point to conditioning the sampling distribution on some features or would have indicated a more efficient mode of sampling.

We have already noted that for binary features assumed to vary randomly between individuals essentially identical conclusions emerge with no special assumption about the sampling procedure but the extremely strong assumption that the population features correspond to independent and identically distributed random variables. We call such an approach *model-based* or equivalently based on a *superpopulation model*. That is, the finite population under study is regarded as itself a sample from a larger universe, usually hypothetical.

A very similar conclusion emerges in the Gaussian case. For this, we may assume $\eta_1, \ldots, \eta_N$ are independently normally distributed with mean μ and variance σ^2. Estimating the finite population mean is essentially equivalent to estimating the mean $\bar{y}^*$ of the unobserved individuals and this is a prediction problem of the type discussed in Section 8.1. That is, the target parameter is an average of a fraction f known exactly, the sample, and an unobserved part, a fraction $1-f$. The predictive pivot is, with σ^2 known, the normally distributed quantity

$$(\bar{y}^* - \bar{y})/(\sigma\sqrt{\{1/n + 1/(N-n)\}}) \tag{9.10}$$

and when σ^2 is unknown an estimated variance is used instead. The pivotal distributions are respectively the standard normal and the Student t distribution with $n - 1$ degrees of freedom. Except for the sharper character of the distributional result, the exact distributional result contrasted with asymptotic normality, this is the same as the design-based inference.

One difference between the two approaches is that in the model-based formulation the choice of statistics arises directly from considerations of sufficiency. In the design-based method the justification for the use of $\bar{Y}$ is partly general plausibility and partly that only linear functions of the observed values have a randomization-based mean that involves m_η. The unweighted average $\bar{y}$ has the smallest variance among all such linear unbiased estimates of m_η.

While the model-based approach is more in line with the discussion in the earlier part of this book, the relative freedom from what in many sampling applications might seem very contrived assumptions has meant that the design-based approach has been the more favoured in most discussions of sampling in the social field, but perhaps rather less so in the natural sciences.

Particularly in more complicated problems, but also to aim for greater theoretical unity, it is natural to try to apply likelihood ideas to the design-based approach. It is, however, unclear how to do this. One approach is to regard $(\eta_1, \ldots, \eta_N)$ as the unknown parameter vector. The likelihood then has the following form. For those s that are observed the likelihood is constant when the η_s in question equals the corresponding Y and zero otherwise and the likelihood does not depend on the unobserved η_s. That is, the likelihood summarizes that the observations are what they are and that there is no information about the unobserved individuals. In a sense this is correct and inevitable. If there is no information whatever either about how the sample was chosen or about the structure of the population no secure conclusion can be drawn beyond the individuals actually sampled; the sample might have been chosen in a highly biased way. Information about the unsampled items can come only from an assumption of population form or from specification of the sampling procedure.

9.2.3 A development

Real sampling problems have many complicating features which we do not address here. To illustrate further aspects of the interplay between design- and model-based analyses it is, however, useful to consider the following extension. Suppose that for each individual there is a further variable z and that these are known for all individuals. The finite population mean of z is denoted by m_z and is thus known. The information about z might indeed be used to set up a modified and more efficient sampling scheme, but we continue to consider random sampling without replacement.

Suppose further that it is reasonable to expect approximate proportionality between the quantity of interest η and z. Most commonly z is some measure of size expected to influence the target variable roughly proportionally. After the sample is taken $\bar{y}$ and the sample mean of z, say $\bar{z}$, are known. If there were exact proportionality between η and z the finite population mean of η would be

$$\tilde{m}_\eta = \bar{y} m_z / \bar{z} \tag{9.11}$$

and it is sensible to consider this as a possible estimate of the finite population mean m_η in which any discrepancy between $\bar{z}$ and m_z is used as a base for a proportional adjustment to $\bar{y}$.

A simple model-based theory of this can be set out in outline as follows. Suppose that the individual values, η_k, and therefore if observed y_k, are random variables of the form

$$\eta_k = \beta z_k + \zeta_k \sqrt{z_k}, \tag{9.12}$$

where $\zeta_1, \ldots, \zeta_N$ are independently normally distributed with zero mean and variance σ_ζ^2. Conformity with this representation can to some extent be tested from the data. When z_k is a measure of size and η_k is an aggregated effect over the individual unit, the square root dependence and approximate normality of the error terms have some theoretical justification via a Central Limit like effect operating within individuals.

Analysis of the corresponding sample values is now possible by the method of weighted least squares or more directly by ordinary least squares applied to the representation

$$y_k/\sqrt{z_k} = \beta\sqrt{z_k} + \zeta_k, \qquad (9.13)$$

leading to the estimate $\hat{\beta} = \bar{y}/\bar{z}$ and to estimates of σ_ζ^2 and $\mathrm{var}(\hat{\beta}) = \sigma_\zeta^2/(n\bar{z})$. Moreover, σ_ζ^2 is estimated by

$$s_\zeta^2 = \sum \frac{(y_k - \hat{\beta} z_k)^2/z_k}{n-1}. \qquad (9.14)$$

The finite population mean of interest, m_η, can be written in the form

$$n\bar{y} + (1-f)\Sigma^* \eta_l/(N-n) = f\bar{y} + (1-f)\{\beta \Sigma^*(z_l + \zeta_l\sqrt{z_l})\} \qquad (9.15)$$

where Σ^* denotes summation over the individuals not sampled. Because

$$\{n\bar{z} + (1-f)\Sigma^* z_l\}/N = m_z, \qquad (9.16)$$

it follows, on replacing β by $\hat{\beta}$ that the appropriate estimate of m_η is $\tilde{m}_\eta$ and that its variance is

$$\frac{(m_z - f\bar{z})m_z}{N\bar{z}}\sigma_\zeta^2. \qquad (9.17)$$

This can be estimated by replacing σ_ζ^2 by s_ζ^2 and hence an exact pivot formed for estimation of m_η.

The design-based analysis is less simple. First, for the choice of $\tilde{m}_\eta$ we rely either on the informal argument given when introducing the estimate above or on the use of an estimate that is optimal under the very special circumstances of the superpopulation model but whose properties are studied purely in a design-based formulation.

For this, the previous discussion of sampling without replacement from a finite population shows that $(\bar{y}, \bar{z})$ has mean (m_y, m_z) and has covariance matrix, in a notation directly extending that used for $v_{\eta\eta}$,

$$\begin{pmatrix} v_{\eta\eta} & v_{\eta z} \\ v_{z\eta} & v_{zz} \end{pmatrix} (1-f)/n. \qquad (9.18)$$

If n is large and the population size N is such that $0 < f < 1$, the design-based sampling distribution of the means is approximately bivariate normal and the conditional distribution of $\bar{y}$ given $\bar{z}$ has approximately the form

$$m_\eta + (v_{\eta z}/v_{zz})(\bar{z} - m_z) + \zeta, \tag{9.19}$$

where ζ has zero mean and variance $v_{\eta\eta \cdot z}(1-f)/n$, and $v_{\eta\eta \cdot z} = v_{\eta\eta} - v_{\eta z}^2/v_{zz}$ is the variance residual to the regression on z. We condition on $\bar{z}$ for the reasons discussed in Chapters 4 and 5. In principle $\bar{z}$ has an exactly known probability distribution although we have used only an asymptotic approximation to it.

Now, on expanding to the first order, the initial estimate is

$$\tilde{m}_\eta = m_\eta + (\bar{y} - m_\eta) - m_\eta(\bar{z} - m_z)/m_z \tag{9.20}$$

and, on using the known form of $\bar{y}$, this gives that

$$\tilde{m}_\eta = m_\eta + (\bar{z} - m_z)(v_{\eta z}/v_{zz} - m_\eta/m_z) + \zeta; \tag{9.21}$$

in the penultimate term we may replace m_η by $\bar{y}$.

Now if the superpopulation model holds $(v_{\eta z}/v_{zz} - m_\eta/m_z) = O_p(1/\sqrt{n})$ and the whole term is $O_p(1/n)$ and can be ignored, but in general this is not the case and the adjusted estimate

$$\tilde{m}'_\eta = \frac{\bar{y} m_z}{\bar{z}} - \frac{(\bar{z} - m_z)\bar{y}}{m_z} \tag{9.22}$$

is appropriate with variance $v_{\eta\eta \cdot z}(1-f)/n$ as indicated. Note that in the unlikely event that η and z are unrelated this recovers $\bar{y}$.

The model-based analysis hinges on the notion, sometimes rather contrived, that the finite population of interest is itself a random sample from a superpopulation. There are then needed strong structural assumptions about the superpopulation, assumptions which can to some extent be tested from the data. There results a clear method of estimation with associated distributional results. By contrast the design-based approach does not on its own specify the appropriate estimate and the associated standard errors and distributional theory are approximations for large sample sizes but assume much less.

9.3 Design of experiments

9.3.1 General remarks

We now turn to the mathematically strongly related but conceptually very different issue of the role of randomization in the design of experiments, i.e., of investigations in which the investigator has in principle total control over the system under study. As a model of the simplest such situation we suppose that there are $2m$ experimental units and two treatments, T and C, to be compared; these can be thought of as a new treatment and a control. The investigator assigns one treatment to each experimental unit and measures a resulting response, y.

We shall introduce and compare two possible designs. In the first, a so-called *completely randomized design*, m of the $2m$ units are selected at random without replacement to receive T and the remainder receive C. For the second design we assume available supplementary information about the units on the basis of which the units can be arranged in pairs. In some aspect related to the response, there should be less variation between units within a pair than between units in different pairs. Then within each pair one unit is selected at random to receive treatment T, the other receiving C. This forms a *matched pair design*, a special case of what for more than two treatments is called a *randomized complete block design*.

We compare a model-based and a design-based analysis of these two designs.

9.3.2 Model-based formulation

A model-based formulation depends on the nature of the response variable, for example on whether it is continuous or discrete or binary and on the kind of distributional assumptions that are reasonable. One fairly general formulation could be based on an exponential family representation, but for simplicity we confine ourselves here largely to normal-theory models.

In the absence of a pairing or blocking structure, using a completely randomized design, the simplest normal-theory model is to write the random variable corresponding to the observation on unit s for $s = 1, \ldots, 2m$ in the symmetrical form

$$Y_s = \mu + d_s \psi + \epsilon_s. \tag{9.23}$$

Here $d_s = 1$ if unit s receives T and $d_s = -1$ if unit s receives C and ϵ_s are independently normally distributed with zero mean and variance σ_{CR}^2. In this formulation the treatment effect is $2\psi = \Delta$, say, assumed constant, and σ_{CR}^2 incorporates all sources of random variation arising from variation between equivalent experimental units and from measurement error.

The corresponding formulation for the matched pair design specifies for $k = 1, \ldots, m; t = T, C$ that the observation on unit k, t is the observed value of the random variable

$$Y_{kt} = \mu + \lambda_k + d_{kt} \psi + \epsilon_{kt}. \tag{9.24}$$

Here again the ϵ are independently normally distributed with zero mean but now in general with a different variance σ_{MP}^2. The general principle involved in writing down this and similar models for more complicated situations is that if some balance is inserted into the design with the objective of improving precision then this balance should be represented in the model, in the present case by including the possibly large number of nuisance parameters λ_k, representing variation between pairs. Note that either design might have been used and in notional comparisons of them the error variances are not likely to be the same. Indeed the intention behind the matched design is normally that σ_{MP}^2 is likely to be appreciably less than σ_{CR}^2.

In general the analysis of the second model will need to recognize the relatively large number of nuisance parameters present, but for a linear least squares analysis this raises no difficulty. In both designs the difference between the two treatment means estimates $\Delta = 2\psi$ with variance $2/m$ times the appropriate variance. This variance is estimated respectively by the appropriate residual mean squares, namely by

$$\{\Sigma_{s \subset T}(Y_s - \bar{Y}_T)^2 + \Sigma_{s \subset C}(Y_s - \bar{Y}_C)^2\}/(2m - 2) \qquad (9.25)$$

and by

$$\{\Sigma(Y_{kT} - Y_{kC})^2 - m(\bar{Y}_T - \bar{Y}_C)^2\}/(m - 1). \qquad (9.26)$$

Here in both formulae $\bar{Y}_T$ and $\bar{Y}_C$ are the means of the observations on T and on C and in the first formula $s \subset T$, for example, refers to a sum over units receiving T.

It can be enlightening, especially in more complicated situations, to summarize such data in analysis of variance tables which highlight the structure of the data as well as showing how to estimate the residual variance.

These arguments lead to pivots for inference about Δ, the pivotal distributions being of the Student t form with respectively $2m - 2$ and $m - 1$ degrees of freedom.

9.3.3 Design-based formulation

For a design-based discussion it is simplest to begin with no explicit block or pair structure and with a null hypothesis of no treatment effect. Consider what may be called the *strong null hypothesis* that the response observed on unit s is a constant, say, η_s, characteristic of the sth unit and is independent of the allocation of treatments to it and other units. That is, the treatments have no differential effect of any sort (in terms of the particular response variable y). Note that this hypothesis has no stochastic component. It implies the existence

9.3 Design of experiments

of two potential physical possibilities for each unit, one corresponding to each treatment, and that only one of these can be realized. Then it is hypothesized that the resulting values of y are identical.

Note that if an independent and identically distributed random variable were added to each possible response, one for T and one for C, so that the difference between T and C for a particular unit became a zero mean random variable identically distributed for each unit, the observable outcomes would be undetectably different from those under the original strong null hypothesis. That is, the nonstochastic character of the strong null hypothesis cannot be tested empirically or at least not without supplementary information, for example about the magnitude of unexplained variation to be anticipated.

The analysis under this null hypothesis can be developed in two rather different ways. One is to note that $\bar{Y}_T$, say, is in the completely randomized design the mean of a random sample of size m drawn randomly without replacement from the finite population of size $2m$ and that $\bar{Y}_C$ now has the role played by $\bar{Y}^*$ in the sampling-theory discussion of the complementary sample. It follows as in the previous section that

$$E_R(\bar{Y}_T - \bar{Y}_C) = 0, \quad \text{var}_R(\bar{Y}_T - \bar{Y}_C) = 2v_\eta/m \qquad (9.27)$$

and that v_η is estimated to a rather close approximation by the model-based estimate of variance. To see the form for the variance of the estimated effect, note that $(\bar{Y}_T + \bar{Y}_C)$ is twice the mean over all units and is a constant in the randomization. Thus the difference of means has the variance of twice that of the sample mean $\bar{Y}_T$, itself a random sample corresponding to a sampling fraction of $1/2$. This argument leads to the same test statistic as in the model-based approach. Combined with an appeal to a form of the Central Limit Theorem, we have an asymptotic version of the model-based analysis of the completely randomized design. We shall see later that an exact distributional analysis is in principle possible.

For the matched pair design the analysis is the same in general approach but different in detail. The most direct formulation is to define X_k, a new variable for each pair, as the difference between the observation on T and that on C. Under the strong null hypothesis X_k is equally likely to be x_k or $-x_k$, where x_k is the observed difference for the kth pair. It follows that under the randomization distribution $\bar{X} = \bar{Y}_T - \bar{Y}_C$ has zero mean and variance

$$\Sigma x_k^2/m. \qquad (9.28)$$

The test statistic is a monotonic function of that involved in the model-based discussion and the tests are asymptotically equivalent.

For comparison with the completely randomized design, note that the variance equivalent to that in the linear model is

$$\Sigma x_k^2/(2m) = \Sigma(\eta_{k1} - \eta_{k0})^2/(2m), \qquad (9.29)$$

as contrasted with what it would be if the completely randomized design were used on the same set of units

$$\Sigma(\eta_{kl} - \bar{\eta}_{..})^2/(2m - 1). \qquad (9.30)$$

A rather different approach is the following. Choose a test statistic for comparing the treatments. This could be, but does not have to be, $\bar{Y}_T - \bar{Y}_C$ or the corresponding standardized test statistic used above. Now under the strong null hypothesis the value of the test statistic can be reconstructed for all possible treatment allocations by appropriate permutation of T and C. Then, because the design gives to all these arrangements the same probability, the exact randomization distribution can be found. This both allows the choice of other test statistics and also avoids the normal approximation involved in the initial discussion.

Table 9.1 illustrates a very simple case of a matched pair experiment. Completion of the enumeration shows that 7 of the 32 configurations give a total summed difference of 8, the observed value, or more and, the distribution being symmetrical about zero, another 7 show a deviation as great as or more extreme than that observed but in the opposite direction. The one-sided p-value

Table 9.1. *Randomization distribution for a simple matched pair experiment. First column gives the observed differences between two treatments. Remaining columns are the first few of the 2^5 configurations obtained by changing signs of differences. Under randomization null hypothesis all 2^5 possibilities have equal probability. Last row gives column totals.*

Observed difference							
6	6	−6	6	−6	6	−6	...
3	3	3	−3	−3	3	3	...
−2	2	2	2	2	−2	−2	...
2	2	2	2	2	2	2	...
−1	−1	−1	−1	−1	−1	−1	...
8	14	2	8	−4	10	−2	...

is $7/32 = 0.22$. A normal approximation based on the known mean and variance and with a continuity correction of the observed difference from -8 to -7 is $\Phi(-7/\sqrt{54}) = 0.17$. Note that because the support of the test statistic is a subset of even integers, the continuity correction is 1 not the more usual $1/2$.

A first point is that in the randomization distribution all possible data configurations have the same value of Σx_k^2 so that, for example, the Student t statistic arising in the model-based approach is a monotone function of the sample mean difference or total. We therefore for simplicity take the sample total Σx_k as the test statistic rather than Student t and form its distribution by giving each possible combination of signs in $\Sigma(\pm x_k)$ equal probability 2^{-m}. The one-sided p-value for testing the strong null hypothesis is thus the proportion of the 2^m configurations with sample totals as large or larger than that observed. Alternatively and more directly, computer enumeration can be used to find the exact significance level, or some form of sampling used to estimate the level with sufficient precision. Improvements on the normal approximation used above can sometimes be usefully found by noting that the test statistic is the sum of independent nonidentically distributed random variables and has the cumulant generating function

$$\log E_R(e^{p\Sigma X_k}) = \Sigma \log \cosh(p x_k) \tag{9.31}$$

from which higher-order cumulants can be found and corrections for nonnormality introduced.

Such tests are called *randomization tests*. They are formally identical to but conceptually quite different from the permutation tests outlined briefly in Chapter 3. Here the basis of the procedure is the randomization used in allocating the treatments; there is no assumption about the stochastic variability of the individual units. By contrast, the validity of the permutation tests hinges on the assumed independence and identical distributional form of the random variability.

In the design-based test the most extreme significance level that could be achieved is 2^{-m} in a one-sided test and 2^{-m+1} in a two-sided test. Thus for quite small m there is inbred caution in the design-based analysis that is not present in the normal-theory discussion. This caution makes qualitative sense in the absence of appreciable prior knowledge about distributional form. There is a further important proviso limiting the value of randomization in isolated very small investigations. We have implicitly assumed that no additional conditioning is needed to make the randomization distribution relevant to the data being analysed. In particular, the design will have taken account of relevant features recognized in advance and for reasonably large m with a large number of arrangements among which to randomize it will often be reasonable to regard these arrangements as essentially equivalent. For very small m, however, each arrangement may have distinctive features which may be particularly relevant to interpretation. The randomization distribution then loses much or all of its force. For example in a very small psychological experiment it might be sensible to use gender as a basis for pairing the individuals, disregarding age as likely to be largely irrelevant. If later it was found that the older subjects fell predominantly in, say, the control group the relevance of the randomization analysis would be suspect. The best procedure in such cases may often be to use the most appealing systematic arrangement. Randomization simply of the names of the treatments ensures some very limited protection against accusations of biased allocation.

Example 9.1. *Two-by-two contingency table.* If the response variable is binary, with outcomes 1 (success) and 0 (failure), the randomization test takes a very simple form under the strong null hypothesis. That is, the total number of 1s, say r, is considered fixed by hypothesis and the randomization used in design ensures that the number of 1s in, say, sample 1 is the number of 1s in a sample of size m drawn randomly without replacement from a finite population of r 1s and $n - r$ 0s. This number has a hypergeometric distribution, recovering Fisher's exact test as a pure randomization test. The restriction to samples of equal size is unnecessary here.

In earlier discussions of the 2×2 table, marginal totals have been fixed either by design or by what was termed technical conditioning. Here the total number of 1s is fixed for a third reason, namely in the light of the strong null hypothesis that the outcome on each study individual is totally unaffected by the treatment allocation.

For large m there is a broad agreement between the randomization analysis and any model-based analysis for a given distributional form and for the choice of test statistic essentially equivalent to Student t. This is clear because the model-based distribution is in effect a mixture of a large number of randomization distributions.

The above discussion has been solely in terms of the null hypothesis of no treatment effect. The simplest modification of the formalism to allow for nonzero effects, for continuous measurements, is to suppose that there is a constant Δ such that on experimental unit s we observe $\eta_s + \Delta$ if T is applied and η_s otherwise.

This is the special case for two treatments of *unit-treatment additivity*. That is, the response observed on a particular unit is the sum of a constant characteristic of the unit and a constant characteristic of the treatment applied to

9.3 Design of experiments

that unit and also is independent of the treatment allocation to other units. It is again a deterministic assumption, testable only indirectly. Confidence limits for Δ are obtained by using the relation between confidence intervals and tests. Take an arbitrary value of Δ, say Δ_0, and subtract it from all the observations on T. Test the new data for consistency with the strong null hypothesis by using the randomization test. Repeat this in principle for all values of Δ_0 and take as confidence set all those values not rejected at the appropriate level.

Adjustment of estimated treatment effects for concomitant variables, in principle measured before randomization to ensure that they are not affected by treatment effects, is achieved in a model-based approach typically by some appropriate linear model. A design-based analysis is usually approximate broadly along the lines sketched in Section 9.2 for sampling.

One of the primary advantages of the design-based approach is that the same formulation of unit-treatment additivity leads for a broad class of randomization patterns, i.e., designs, to an unbiased estimate of the treatment differences and to an unbiased estimate of the effective error variance involved. This contrasts with the model-based approach which appears to make a special ad hoc assumption for each design. For example for a Latin square design a special so-called main effect model that appears to be a highly specific assumption is postulated. This leads to the estimation of error via a particular residual sum of squares. In fact the same residual sum of squares arises in the design-based analysis directly from the specific randomization used and without any special assumption.

There is the following further conceptual difference between the two approaches. A model-based approach might be regarded as introducing an element of extrapolation in that the conclusions apply to some ensemble of real or hypothetical repetitions that are involved in defining the parameters. The design-based approach, by contrast, is concerned with elucidating what happened in the particular circumstances of the investigation being analysed; how likely is it that the adverse play of chance in the way the treatments were allocated in the study has distorted the conclusions seriously? Any question of extrapolation is then one of general scientific principle and method and not a specifically statistical issue.

This discussion of randomization has been from a frequentist standpoint. While the role of randomization in a formal Bayesian theory is not so clear, from one point of view it can be regarded as an informal justification of the independence assumptions built into typical forms of likelihood used in Bayesian inference. This is essentially parallel to the interplay between design- and model-based analyses in frequentist theory. In the more personalistic approaches it might be argued that what is required is that You believe the allocation to be essentially random but empirical experience suggests that, in some contexts at least, this is a very hazardous argument!

Notes 9

Section 9.1. Randomization has three roles in applications: as a device for eliminating biases, for example from unobserved explanatory variables and selection effects; as a basis for estimating standard errors; and as a foundation for formally exact significance tests. The relative importance of these varies greatly between fields. The emphasis in the present discussion is on the conceptual possibility of statistical inference not based on a probabilistic model of the underlying data-generating process. It is important to distinguish permutation tests from the numerically equivalent randomization tests. The former are based on some symmetries induced by the probabilistic model, whereas the latter are a by-product of the procedure used in design. Modern computational developments make the use of permutation tests feasible even in quite complicated problems (Manly, 1997). They provide protection of significance tests against departures of distributional form but not against the sometimes more treacherous failure of independence assumptions.

Section 9.2. General accounts of sampling theory are given by Cochran (1977) and Thompson (1992) and a more mathematical discussion by Thompson (1997). There has been extensive discussion of the relative merits of model- and design-based approaches. For the applications to stereology, including a discussion of model- versus design-based formulations, see Baddeley and Jensen (2005). For a comparison of methods of estimating the variance of ratio estimates, see Sundberg (1994).

Section 9.3. The basic principles of statistical design of experiments were set out by Fisher (1935a) and developed, especially in the context of agricultural field trials, by Yates (1937). For an account of the theory emphasizing the generality of the applications, see Cox and Reid (2000). Fisher introduced randomization and randomization tests but his attitude to the use of the tests in applications is unclear; Yates emphasized the importance of randomization in determining the appropriate estimate of error in complex designs but dismissed the detailed randomization analysis as inferior to least squares. Kempthorne (1952) took the opposing view, regarding least squares analyses as computationally convenient approximations to the randomization analyses. The development from the counterfactual model of experimental design to address studies of causation in observational studies has been extensively studied; see, in particular, Rubin (1978, 1984). By adopting a Bayesian formulation and assigning a prior distribution to the constants characterizing the observational units, the analysis is brought within the standard Bayesian specification.

Appendix A
A brief history

A very thorough account of the history of the more mathematical side of statistics up to the 1930's is given by Hald (1990, 1998, 2006). Stigler (1990) gives a broader perspective and Heyde and Seneta (2001) have edited a series of vignettes of prominent statisticians born before 1900.

Many of the great eighteenth and early nineteenth century mathematicians had some interest in statistics, often in connection with the analysis of astronomical data. Laplace (1749–1827) made extensive use of flat priors and what was then called the method of inverse probability, now usually called Bayesian methods. Gauss (1777–1855) used both this and essentially frequentist ideas, in particular in his development of least squares methods of estimation. Flat priors were strongly criticized by the Irish algebraist Boole (1815–1864) and by later Victorian mathematicians and these criticisms were repeated by Todhunter (1865) in his influential history of probability. Karl Pearson (1857–1936) began as, among other things, an expert in the theory of elasticity, and brought Todhunter's history of that theory to posthumous publication (Todhunter, 1886, 1893).

In one sense the modern era of statistics started with Pearson's (1900) development of the chi-squared goodness of fit test. He assessed this without comment by calculating and tabulating the tail area of the distribution. Pearson had some interest in Bayesian ideas but seems to have regarded prior distributions as essentially frequency distributions. Pearson's influence was dominant in the period up to 1914, his astonishing energy being shown in a very wide range of applications of statistical methods, as well as by theoretical contributions.

While he continued to be active until his death, Pearson's influence waned and it was R. A. Fisher (1890–1962) who, in a number of papers of the highest originality (Fisher, 1922, 1925a, 1934, 1935b), laid the foundations for much of the subject as now known, certainly for most of the ideas stressed in the present book. He distinguished problems of specification, of estimation and of distribution. He introduced notions of likelihood, maximum likelihood and conditional inference. He also had a distinctive view of probability. In parallel with his theoretical work he suggested major new methods of analysis and of experimental design, set out in two books of high impact (Fisher, 1925b, 1935a). He was firmly anti-Bayesian in the absence of specific reasons for adopting a prior distribution (Fisher, 1935a, 1956). In addition to his central role is statistical thinking, he was an outstanding geneticist, in particular being a pioneer in putting the theory of natural selection on a quantitative footing.

Fisher had little sympathy for what he regarded as the pedanticism of precise mathematical formulation and, only partly for that reason, his papers are not always easy to understand. In the mid 1920s J. Neyman (1894–1981), then in Warsaw, and E. S. Pearson (1885–1980) began an influential collaboration initially designed primarily, it would seem, to clarify Fisher's writing. This led to their theory of testing hypotheses and to Neyman's development of confidence intervals, aiming to clarify Fisher's idea of fiducial intervals. As late as 1932 Fisher was writing to Neyman encouragingly about this work, but relations soured, notably when Fisher greatly disapproved of a paper of Neyman's on experimental design and no doubt partly because their being in the same building at University College London brought them too close to one another! Neyman went to Berkeley at the start of World War II in 1939 and had a very major influence on US statistics. The differences between Fisher and Neyman were real, centring, for example, on the nature of probability and the role, if any, for conditioning, but, I think, were not nearly as great as the asperity of the arguments between them might suggest.

Although some of E. S. Pearson's first work had been on a frequency-based view of Bayesian inference, Neyman and Pearson, except for an early elegant paper (Neyman and Pearson, 1933), showed little interest in Bayesian approaches. Wald (1900–1950) formulated a view of statistical analysis and theory that was strongly decision-oriented; he assumed the availability of a utility or loss function but not of a prior distribution and this approach too attracted much attention for a while, primarily in the USA.

Fisher in some of his writing emphasized direct use of the likelihood as a measure of the relation between data and a model. This theme was developed by Edwards (1992) and by Royall (1997) and in much work, not always published, by G. A. Barnard (1915–2002); see, in particular, Barnard et al. (1962).

As explained above, twentieth century interest in what was earlier called the inverse probability approach to inference was not strong until about 1950. The economist John Maynard Keynes (1883–1946) had written a thesis on the objectivist view of these issues but the main contribution to that approach, and a highly influential one, was that of the geophysicist and applied mathematician

Harold Jeffreys (1891–1989). He wrote on the objective degree of belief view of probability in the early 1930s but his work is best approached from the very influential *Theory of Probability*, whose first edition was published in 1939. He justified improper priors, gave rules for assigning priors under some circumstances and developed his ideas to the point where application could be made. Comparable developments in the philosophy of science, notably by Carnap, do not seem to have approached the point of applicability. The most notable recent work on standardized priors as a base for comparative analyses is that of Bernardo (1979, 2005); this aims to find priors which maximize the contribution of the data to the posterior.

At the time of these earlier developments the philosopher F. P. Ramsey (1903–1930) developed a theory of personalistic probability strongly tied to personal decision-making, i.e., linking probability and utility intimately. Ramsey died before developing these and other foundational ideas. The systematic development came independently from the work of de Finetti (1906–1985) and Savage (1917–1971), both of whom linked personalistic probability with decision-making. The pioneering work of I. J. Good (1916–) was also important. There were later contributions by many others, notably de Groot (1931–1989) and Lindley (1923–). There followed a period of claims that the arguments for this personalistic theory were so persuasive that anything to any extent inconsistent with that theory should be discarded. For the last 15 years or so, i.e., since about 1990, interest has focused instead on applications, especially encouraged by the availability of software for Markov chain Monte Carlo calculations, in particular on models of broadly hierarchical type. Many, but not all, of these applications make no essential use of the more controversial ideas on personalistic probability and many can be regarded as having at least approximately a frequentist justification.

The later parts of this brief history may seem disturbingly Anglo-centric. There have been, of course, many individuals making important original contributions in other languages but their impact has been largely isolated.

Appendix B
A personal view

Much of this book has involved an interplay between broadly frequentist discussion and a Bayesian approach, the latter usually involving a wider notion of the idea of probability. In many, but by no means all, situations numerically similar answers can be obtained from the two routes. Both approaches occur so widely in the current literature that it is important to appreciate the relation between them and for that reason the book has attempted a relatively dispassionate assessment.

This appendix is, by contrast, a more personal statement. Whatever the approach to formal inference, formalization of the research question as being concerned with aspects of a specified kind of probability model is clearly of critical importance. It translates a subject-matter question into a formal statistical question and that translation must be reasonably faithful and, as far as is feasible, the consistency of the model with the data must be checked. How this translation from subject-matter problem to statistical model is done is often the most critical part of an analysis. Furthermore, all formal representations of the process of analysis and its justification are at best idealized models of an often complex chain of argument.

Frequentist analyses are based on a simple and powerful unifying principle. The implications of data are examined using measuring techniques such as confidence limits and significance tests calibrated, as are other measuring instruments, indirectly by the hypothetical consequences of their repeated use. In particular, they use the notion that consistency in a certain respect of the data with a specified value of the parameter of interest can be assessed via a p-value. This leads, in particular, to use of sets of confidence intervals, often conveniently expressed in pivotal form, constituting all parameter values consistent with the data at specified levels. In some contexts the emphasis is rather on those possible explanations that can reasonably be regarded as refuted by the data.

The well-known definition of a statistician as someone whose aim in life is to be wrong in exactly 5 per cent of everything they do misses its target.

The objective is to recognize explicitly the possibility of error and to use that recognition to calibrate significance tests and confidence intervals as an aid to interpretation. This is to provide a link with the real underlying system, as represented by the probabilistic model of the data-generating process. This is the role of such methods in analysis and as an aid to interpretation. The more formalized use of such methods, using preset levels of significance, is also important, for example in connection with the operation of regulatory agencies.

In principle, the information in the data is split into two parts, one to assess the unknown parameters of interest and the other for model criticism. Difficulties with the approach are partly technical in evaluating p-values in complicated systems and also lie in ensuring that the hypothetical long run used in calibration is relevant to the specific data under analysis, often taking due account of how the data were obtained. The choice of the appropriate set of hypothetical repetitions is in principle fundamental, although in practice much less often a focus of immediate concern. The approach explicitly recognizes that drawing conclusions is error-prone. Combination with qualitative aspects and external knowledge is left to a qualitative synthesis. The approach has many advantages, not least in its ability, for example especially in the formulation of Neyman and Pearson, to provide a basis for assessing methods of analysis put forward on informal grounds and for comparing alternative proposals for data collection and analysis. Further, it seems clear that any proposed method of analysis that in repeated application would mostly give misleading answers is fatally flawed.

Some versions of the Bayesian approach address the same issues of interpretation by a direct use of probability as an impersonal measure of uncertainty, largely ignoring other sources of information by the use of priors that are in some sense standardized or serving as reference points. While there is interesting work on how such reference priors may be chosen, it is hard to see a conceptual justification for them other than that, at least in low-dimensional problems, they lead, at least approximately, to procedures with acceptable frequentist properties. In particular, relatively flat priors should be used as such with caution, if at all, and regarded primarily as a possible route to reasonable confidence limits. Flat priors in multidimensional parameters may lead to absurd answers. From this general perspective one view of Bayesian procedures is that, formulated carefully, they may provide a convenient algorithm for producing procedures that may have very good frequentist properties.

A conceptually entirely different Bayesian approach tackles the more ambitious task of including sources of information additional to the data by doing this not qualitatively but rather quantitatively by a specially chosen prior distribution. This is, of course, sometimes appealing. In particular, the interpretation of data often involves serious sources of uncertainty, such as systematic errors, other than those directly represented in the statistical model used to represent the data-generating process, as well as uncertainties about the model itself. In a more positive spirit, there may be sources of information additional to the data under analysis which it would be constructive to include. The central notion, used especially in the task of embracing all sorts of uncertainty, is that probability assessments should be coherent, i.e., *internally* consistent. This leads to the uncertainty attached to any unsure event or proposition being measured by a single real number, obeying the laws of probability theory. Any such use of probability as representing a notion of degree of belief is always heavily conditional on specification of the problem.

Major difficulties with this are first that any particular numerical probability, say $1/2$, is given exactly the same meaning whether it is Your current assessment based on virtually no information or it is solidly based on data and theory. This confusion is unacceptable for many interpretative purposes. It raises the rhetorical question: is it really sensible to suppose that uncertainty in all its aspects can always be captured in a single real number?

Next, the issue of temporal coherency discussed in Section 5.10 is often largely ignored; obtaining the data, or preliminary analysis, may entirely properly lead to reassessment of the prior information, undermining many of the conclusions emanating from a direct application of Bayes' theorem. Direct uses of Bayes' theorem are in effect the combination of different sources of information without checking for mutual consistency. Issues of model criticism, especially a search for ill-specified departures from the initial model, are somewhat less easily addressed within the Bayesian formulation.

Finally, perhaps more controversially, the underlying definition of probability in terms of Your hypothetical behaviour in uncertain situations, subject only to internal consistency, is in conflict with the primacy of considering evidence from the real world. That is, although internal consistency is desirable, to regard it as overwhelmingly predominant is in principle to accept a situation of always being self-consistently wrong as preferable to some inconsistent procedure that is sometimes, or even quite often, right.

This is not a rejection of the personalistic Bayesian approach as such; that may, subject to suitable sensitivity analysis, sometimes provide a fruitful way of injecting further information into an analysis or using data in a more speculative way, especially, but, not only, as a basis for personal decision-making. Rather it is an emphatic rejection of the notion that the axioms of personalistic probability are so compelling that methods not explicitly or implicitly using that approach are to be rejected. Some assurance of being somewhere near the truth takes precedence over internal consistency.

One possible role for a prior is the inclusion of expert opinion. This opinion may essentially amount to a large number of empirical data informally analysed. There are many situations where it would be unwise to ignore such information, although there may be some danger of confusing well-based information with the settling of issues by appeal to authority; also empirical experience may suggest that expert opinion that is not reasonably firmly evidence-based may be forcibly expressed but is in fact fragile. The frequentist approach does not ignore such evidence but separates it from the immediate analysis of the specific data under consideration.

The above remarks address the use of statistical methods for the analysis and interpretation of data and for the summarization of evidence for future use. For automatized decision-making or for the systemization of private opinion, the position is rather different. Subject to checks of the quality of the evidence-base, use of a prior distribution for relevant features may well be the best route for ensuring that valuable information is used in a decision procedure. If that information is a statistical frequency the Bayesian procedure is uncontroversial, subject to the mutual consistency of the information involved. It is significant that the major theoretical accounts of personalistic probability all emphasize its connection with decision-making. An assessment of the evidence-base for the prior remains important.

A further type of application often called Bayesian involves risk assessment, usually the calculation of the probability of an unlikely event arising from a concatenation of circumstances, themselves with poorly evaluated probabilities. Such calculations can certainly be valuable, for example in isolating critical pathways to disaster. They are probably usually best regarded as frequentist probability calculations using poorly known elements and as such demand careful examination of dependence both on the values of marginal probabilities and on the strong, if not always explicit, independence assumptions that often underlie such risk assessments. There are standard techniques of experimental design to help with the sensitivity analyses that are desirable in such work.

The whole of this appendix, and indeed the whole book, is concerned with *statistical* inference. The object is to provide ideas and methods for the critical analysis and, as far as feasible, the interpretation of empirical data arising from a single experimental or observational study or from a collection of broadly similar studies bearing on a common target. The extremely challenging issues of *scientific* inference may be regarded as those of synthesising very different kinds of conclusions if possible into a coherent whole or theory and of placing specific analyses and conclusions within that framework. This process is surely subject to uncertainties beyond those of the component pieces of information and, like statistical inference, has the features of demanding stringent evaluation of the consistency of information with proposed explanations. The use, if any, in this process of simple *quantitative* notions of probability and their numerical assessment is unclear and certainly outside the scope of the present discussion.

References

Akahira, M. and Takeuchi, K. (2003). *Joint Statistical Papers of Akahira and Takeuchi.* Singapore: World University Press.
Aitchison, J. and Dunsmore, I. R. (1975). *Statistical Prediction Analysis.* Cambridge: Cambridge University Press.
Aitchison, J. and Silvey, S. D. (1958). Maximum likelihood estimation of parameters subject to restraints. *Ann. Math. Statist.* **29**, 813–828.
Amari, S. (1985). *Differential-geometric Methods in Statistics.* Lecture Notes in Statistics, Vol. 28. New York: Springer.
Andersen, P. K., Borgan, Ø., Gill, R. D. and Keiding, N. (1992). *Statistical Models Based on Counting Processes.* New York: Springer.
Anderson, T. W. (1973). Asymptotically efficient estimation of covariance matrices with linear structure. *Ann. Statist.* **1**, 135–141.
Anderson, T. W. (2004). *Introduction to Multivariate Analysis.* Third edition. New York: Wiley.
Anscombe, F. J. (1949). Large sample theory of sequential estimation. *Biometrika* **36**, 455–458.
Anscombe, F. J. (1953). Sequential estimation (with discussion). *J. R. Statist. Soc.* B **15**, 1–29.
Azzalini, A. (1983). Maximum likelihood estimation of order m for stationary stochastic processes. *Biometrika* **70**, 381–387.
Azzalini, A. (1996). *Statistical Inference.* London: Chapman and Hall.
Baddeley, A. and Jensen, E. B. V. (2005). *Stereology for Statisticians.* Boca Raton: Chapman and Hall/CRC.
Barnard, G. A., Jenkins, G. M. and Winsten, C. B. (1962). Likelihood inference and time series (with discussion). *J. R. Statist. Soc.* A **125**, 321–372.
Barndorff-Nielsen, O. E. (1978). *Information and Exponential Families in Statistical Theory.* Chichester: Wiley.
Barndorff-Nielsen, O. E. and Cox, D. R. (1984). The effect of sampling rules on likelihood statistics. *Int. Statist. Rev.* **52**, 309–326.
Barndorff-Nielsen, O. E. and Cox, D. R. (1994). *Inference and Asymptotics.* London: Chapman and Hall.
Barndorff-Nielsen, O. E., Cox, D. R. and Reid, N. (1986). The role of differential geometry in statistical theory. *Int. Statist. Rev.* **54**, 83–96.

Barnett, V. and Barnett, V. D. (1999). *Comparative Statistical Inference*. Third edition. Chichester: Wiley.

Bartlett, M. S. (1937). Properties of sufficiency and statistical tests. *Proc. Roy. Soc.* A **160**, 268–282.

Bartlett, M. S. (1953a). Approximate confidence intervals, I. *Biometrika* **40**, 12–19.

Bartlett, M. S. (1953b). Approximate confidence intervals, II. *Biometrika* **40**, 306–317.

Bartlett, M. S. (1957). A note on D. V. Lindley's statistical paradox. *Biometrika* **44**, 533–534.

Berger, J. (2004). The case for objective Bayesian analysis. *Bayesian Analysis* **1**, 1–17.

Bernardo, J. M. (1979). Reference posterior distributions for Bayesian inference (with discussion). *J. R. Statist. Soc.* B **41**, 113–147.

Bernardo, J. M. (2005). Reference analysis. *Handbook of Statistics*. Vol. 35. Amsterdam: Elsevier.

Bernardo, J. M. and Smith, A. F. M. (1994). *Bayesian Theory*. Chichester: Wiley.

Besag, J. E. (1974). Spatial interaction and the statistical analysis of lattice systems (with discussion). *J. R. Statist. Soc.* B **36**, 192–236.

Besag, J. E. (1977). Efficiency of pseudo-likelihood estimates for simple Gaussian fields. *Biometrika* **64**, 616–618.

Besag, J. E. and Mondal, D. (2005). First-order intrinsic autoregressions and the de Wijs process. *Biometrika* **92**, 909–920.

Birnbaum, A. (1962). On the foundations of statistical inference (with discussion). *J. Am. Statist. Assoc.* **57**, 269–326.

Birnbaum, A. (1969). Concepts of statistical evidence. In *Philosophy, Science and Method: Essays in Honor of E. Nagel*, pp. 112–143. New York: St Martin's Press.

Box, G. E. P. (1980). Sampling and Bayes inference in scientific inference and robustness (with discussion). *J. R. Statist. Soc.* A **143**, 383–430.

Box, G. E. P. and Tiao, G. C. (1973). *Bayesian Inference in Statistical Analysis*. Reading, Mass: Addison Wesley.

Brazzale, A. R., Davison, A. C. and Reid, N. (2007). *Applied Asymptotics*. Cambridge: Cambridge University Press.

Breslow, N. E. and Day, N. E. (1980). *The Design and Analysis of Case-Control Studies*. Lyon: IARC.

Brockwell, P. J. and Davis, R. A. (1991). *Time Series: Theory and Methods*. New York: Springer.

Brockwell, P. J. and Davis, R. A. (1998). *Time Series: Theory and Methods*. Vol. 2. New York: Springer.

Butler, R. W. (1986). Predictive likelihood inference with applications (with discussion). *J. R. Statist. Soc.* B **48**, 1–38.

Butler, R. W. (2007). *Saddlepoint Approximations with Applications*. Cambridge: Cambridge University Press.

Casella, G. and Berger, R. L. (1990). *Statistical Inference*. Pacific Grove: Wadsworth and Brooks/Cole.

Christensen, R. and Utts, J. (1992). Bayesian resolution of the exchange paradox. *Amer. Statistician* **46**, 274–296.

Cochran, W. G. (1937). The efficiencies of the binomial series tests of significance of a mean and of a correlation coefficient. *J. R. Statist. Soc.* **100**, 69–73.

Cochran, W. G. (1977). *Sampling Techniques*. New York: Wiley.

Copas, J. and Eguchi, S. (2005). Local model uncertainty and incomplete-data bias (with discussion). *J. R. Statist. Soc.* B **67**, 459–513.
Cox, D. R. (1958a). The regression analysis of binary sequences (with discussion). *J. R. Statist. Soc.* B **20**, 215–242.
Cox, D. R. (1958b). Two further applications of a model for binary regression. *Biometrika* **45**, 562–565.
Cox, D. R. (1958c). Some problems connected with statistical inference. *Ann. Math. Statist.* **29**, 357–372.
Cox, D. R. (1961). Tests of separate families of hypotheses. *Proc. 4th Berkeley Symp.* **1**, 105–123.
Cox, D. R. (1962). Further results on tests of separate families of hypotheses. *J. R. Statist. Soc.* B **24**, 406–424.
Cox, D. R. (1972). Regression models and life-tables (with discussion) *J. R. Statist. Soc.* B **34**, 187–220.
Cox, D. R. (1975). Partial likelihood. *Biometrika* **62**, 269–276.
Cox, D. R. (1977). The role of significance tests (with discussion). *Scand. J. Statist.* **4**, 49–70.
Cox, D. R. and Hinkley, D. V. (1974). *Theoretical Statistics.* London: Chapman and Hall.
Cox, D. R. and Medley, G. F. (1989). A process of events with notification delay and the forecasting of AIDS. *Phil. Trans. Roy. Soc.* B **325**, 135–145.
Cox, D. R. and Reid, N. (2000). *The Theory of the Design of Experiments.* Boca Raton: Chapman and Hall/CRC.
Cox, D. R. and Reid, N. (2004). A note on pseudolikelihood constructed from marginal densities. *Biometrika* **91**, 729–737.
Cox, D. R. and Snell, E. J. (1988). *The Analysis of Binary Data.* Second edition. London: Chapman and Hall.
Cox, D. R. and Solomon, P. J. (2002). *Components of Variance.* Boca Raton: Chapman and Hall/CRC.
Cox, D. R. and Wermuth, N. (1990). An approximation to maximum likelihood estimates in reduced models. *Biometrika* **77**, 747–761.
Creasy, M. A. (1954). Limits for the ratio of means (with discussion). *J. R. Statist. Soc.* B **16**, 186–194.
Daniels, H. E. (1954). Saddlepoint approximations in statistics. *Ann. Math. Statist.* **25**, 631–650.
Davies, R. B. (1977). Hypothesis testing when a nuisance parameter is present only under the alternative. *Biometrika* **64**, 247–254.
Davies, R. B. (1987). Hypothesis testing when a nuisance parameter is present only under the alternative. *Biometrika* **74**, 33–43.
Davison, A. C. (2003). *Statistical Models.* Cambridge: Cambridge University Press.
Davison, A. C. and Hinkley, D.V. (1997). *Bootstrap Methods and Their Application.* Cambridge: Cambridge University Press.
Dawid, A. P. (1975). Comment on the paper by Efron (1975). *Ann. Statist.* **3**, 1231–1234.
Dempster, A. P. (1972). Covariance selection. *Biometrics* **28**, 157–175.
Dempster, A. P., Laird, N. M. and Rubin, D. B. (1977). Maximum likelihood estimation from incomplete data via the EM algorithm (with discussion). *J. R. Statist. Soc.* B **39**, 1–38.

Edwards, A. W. F. (1992). *Likelihood*. Second edition. Cambridge: Cambridge University Press.
Efron, B. (1979). Bootstrap methods: another look at the jackknife. *Ann. Statist.* **7**, 1–26.
Efron, B. and Hinkley, D. V. (1978). Assessing the accuracy of the maximum likelihood estimator: observed versus expected Fisher information. *Biometrika* **65**, 457–487.
Farewell, V. (1979). Some results on the estimation of logistic models based on retrospective data. *Biometrika* **66**, 27–32.
Fisher, R. A. (1922). On the mathematical foundations of theoretical statistics. *Phil. Trans. Roy. Soc.* A **222**, 309–368.
Fisher, R. A. (1925a). Theory of statistical estimation. *Proc. Camb. Phil. Soc.* **22**, 700–725.
Fisher, R. A. (1925b). *Statistical Methods for Research Workers*. Edinburgh: Oliver and Boyd. And subsequent editions
Fisher, R. A. (1934). Two new properties of maximum likelihood. *Proc. Roy. Soc.* A **144**, 285–307.
Fisher, R. A. (1935a). *Design of Experiments*. Edinburgh: Oliver and Boyd. And subsequent editions.
Fisher, R. A. (1935b). The logic of inductive inference (with discussion). *J. R. Statist. Soc.* **98**, 39–54.
Fisher, R. A. (1936). The use of multiple measurements in taxonomic problems. *Ann. Eugenics* **7**. 179–188.
Fisher, R. A. (1940). The precision of discriminant functions. *Ann. Eugenics*. **10**, 422–429.
Fisher, R. A. (1950). The significance of deviations from expectations in a Poisson series. *Biometrics* **6**, 17–24.
Fisher, R. A. (1956). *Statistical Methods and Scientific Inference*. Edinburgh: Oliver and Boyd.
Fraser, D. A. S. (1979). *Inference and Linear Models*. New York: McGraw Hill.
Fraser, D. A. S. (2004). Ancillaries and conditional inference (with discussion). *Statist. Sci.* **19**, 333–369.
Garthwaite, P. H., Kadane, J. B. and O'Hagan, A. (2005). Statistical methods for elicitating probability distributions. *J. Amer. Statist. Assoc.* **100**, 680–700.
Geisser, S. (1993). *Predictive Inference: An Introduction*. London: Chapman and Hall.
Godambe, V.P., editor (1991). *Estimating Functions*. Oxford: Oxford University Press.
Green, P. J. and Silverman, B. W. (1994). *Nonparametric Regression and Generalized Linear Models*. London: Chapman and Hall.
Greenland, S. (2005). Multiple-bias modelling for analysis of observational data (with discussion). *J. R. Statist. Soc.* A **168**, 267–306.
Hacking, I. (1965). *Logic of Statistical Inference*. Cambridge: Cambridge University Press.
Hald, A. (1990). *A History of Probability and Statistics and Their Applications before 1750*. New York: Wiley.
Hald, A. (1998). *A History of Mathematical Statistics from 1750 to 1930*. New York: Wiley.
Hald, A. (2006). *A History of Parametric Statistical Inference from Bernoulli to Fisher, 1713–1935*. New York: Springer.

Hall, P. and Wang, J. Z. (2005). Bayesian likelihood methods for estimating the end point of a distribution. *J. R. Statist. Soc.* B **67**, 717–729.
Halmos, P. R. and Savage, L. J. (1949). Application of the Radon–Nikodym theorem to the theory of sufficient statistics. *Ann. Math. Statist.* **20**, 225–241.
Heyde, C. C. and Seneta, E. (2001). *Statisticians of the Centuries.* New York: Springer.
Hochberg, J. and Tamhane, A. (1987). *Multiple Comparison Procedures.* New York: Wiley.
Jaynes, E. T. (1976). Confidence intervals versus Bayesian intervals (with discussion). In *Foundations of Probability Theory, Statistical Inference and Statistical Theories of Science*, W. L. Harper and C. A. Hooker, editors, pp. 175–257, Vol. 2, Dordrecht: Reidel.
Jeffreys, H. (1939, 1961). *Theory of Probability.* First edition, 1939, third edition, 1961. Oxford: Oxford University Press.
Jiang, W. and Turnbull, B. (2004). The indirect method: inference based on intermediate statistics – a synthesis and examples. *Statist. Sci.* **19**, 239–263.
Johnson, V. E. (2005). Bayes factors based on test statistics. *J. R. Statist. Soc.* B **67**, 689–701.
Kalbfleisch, J. D. (1975). Sufficiency and conditionality. *Biometrika* **62**, 251–268.
Kalbfleisch, J. D. and Sprott, D. A. (1970). Application of likelihood methods to models involving large numbers of parameters (with discussion). *J. R. Statist. Soc.* B **32**, 175–208.
Kass, R. E. and Raftery, A. E. (1995). Bayes factors. *J. Am. Statist. Assoc.* **90**, 773–795.
Kass, R. E. and Wasserman, L. (1996). The selection of prior probabilities by formal rules. *J. Am. Statist. Assoc.* **91**, 1343–1370.
Kempthorne, O. (1952). *Design of Experiments.* New York: Wiley.
Lange, K. (2000). *Numerical Analysis for Statisticians.* New York: Springer.
Lawless, J. F. and Fridette, M. (2005). Frequentist prediction intervals and predictive distributions. *Biometrika* **92**, 529–542.
Lee, Y. and Nelder, J. A. (2006). Double hierarchical generalized linear models (with discussion). *Appl. Statist.* **55**, 139–186.
Lehmann, E. L. (1998). *Nonparametrics: Statistical Methods and Research.* New York: Wiley.
Lehmann, E. L. and Casella, G. C. (2001). *Theory of Point Estimation.* New York: Springer.
Lehmann, E. L. and Romano, J. P. (2004). *Testing of Statistical Hypotheses.* New York: Springer.
Liang, K. Y. and Zeger, S. L. (1986). Longitudinal data analysis using generalized linear models. *Biometrika* **73**, 13–22.
Lindley, D. V. (1956). On a measure of the information provided by an experiment. *Ann. Math. Statist.* **27**, 986–1005.
Lindley, D. V. (1957). A statistical paradox. *Biometrika* **44**, 187–192.
Lindley, D. V. (1958). Fiducial distributions and Bayes' theorem. *J. R. Statist. Soc.* B **20**, 102–107.
Lindley, D. V. (1990). The present position in Bayesian statistics (with discussion). *Statist. Sci.* **5**, 44–89.

Lindsay, B. (1988). Composite likelihood methods. In *Statistical Inference from Stochastic Processes*, pp. 221–239. Providence, RI: American Mathematical Society.
Liu, J. S. (2002). *Monte Carlo Strategies in Statistical Computing*. New York: Springer.
Manly, B. (1997). *Randomization, Bootstrap and Monte Carlo Methods in Biology*. Boca Raton: Chapman and Hall.
Mayo, D. G. (1996). *Error and the Growth of Experimental Knowledge*. Chicago: University of Chicago Press.
McCullagh, P. (1983). Quasi-likelihood functions. *Ann. Statist.* **11**, 59–67.
McCullagh, P. and Cox, D. R. (1986). Invariants and likelihood ratio statistics. *Ann. Statist.* **14**, 1419–1430.
Meng, X-L. and van Dyk, D. (1997). The EM algorithm – an old party song to a fast new tune (with discussion). *J. R. Statist. Soc.* B **59**, 511–567.
Mitchell, A. F. S. (1967). Discussion of paper by I. J. Good. *J. R. Statist. Soc.* B **29**, 423–424.
Murray, M. K. and Rice, J. W. (1993). *Differential Geometry and Statistics*. London: Chapman and Hall.
Mykland, P. A. (1995). Dual likelihood. *Ann. Statist.* **23**, 396–421.
Nelder, J. A. and Mead, R. (1965). A simplex method for function minimization. *Computer J.* **7**, 308–313.
Neyman, J. and Pearson, E. S. (1933). The testing of statistical hypotheses in relation to probabilities *a priori*. *Proc. Camb. Phil. Soc.* **24**, 492–510.
Neyman, J. and Pearson, E. S. (1967). *Joint Statistical Papers of J. Neyman and E. S. Pearson*. Cambridge: Cambridge University Press and Biometrika Trust.
Owen, A. B. (2001). *Empirical Likelihood*. Boca Raton: Chapman and Hall/CRC.
Pawitan, Y. (2000). *In All Likelihood*. Oxford: Oxford University Press.
Pearson, E. S. (1947). The choice of statistical tests illustrated on the interpretation of data classed in a 2 × 2 table. *Biometrika* **34**, 139–167.
Pearson, K. (1900). On the criterion that a given system of deviations from the probable in the case of a correlated system of variables is such that it can reasonably be supposed to have arisen from random sampling. *Phil. Mag.* **50**, 157–172.
Prentice, R. L. and Pyke, R. (1979). Logistic disease incidence models and case-control studies. *Biometrika* **66**, 403–411.
Rao, C. R. (1973). *Linear Statistical Inference*. Second edition. New York: Wiley.
Reid, N. (2003). Asymptotics and the theory of inference. *Ann. Statist.* **31**, 1695–2095.
Rice, J. A. (1988). *Mathematical Statistics and Data Analysis*. Pacific Grove: Wadsworth and Brooks/Cole.
Ripley, B. D. (1987). *Stochastic Simulation*. New York: Wiley.
Ritz, C. and Skovgaard, I. M. (2005). Likelihood ratio tests in curved exponential families with nuisance parameters present only under the alternative. *Biometrika* **92**, 507–517.
Robert, C. P. and Casella, G. (2004). *Monte Carlo Statistical Methods*. New York: Springer.
Robinson, G. K. (1979). Conditional properties of statistical procedures. *Ann. Statist.* **7**, 742–755.
Ross, G. J. S. (1990). *Nonlinear Estimation*. New York: Springer.
Rotnitzky, A., Cox, D. R., Robins, J. E. and Botthai, H. (2000). Likelihood-based inference with singular information matrix. *Bernoulli* **6**, 243–284.

Royall, R. M. (1997). *Statistical Evidence: A Likelihood Paradigm.* London: Chapman and Hall.
Rubin, D. B. (1978). Bayesian inference for causal effects: the role of randomization. *Ann. Statist.* **6**, 35–48.
Rubin, D. B. (1984). Bayesianly justifiable and relevant frequency calculations for the applied statistician. *Ann. Statist.* **12**, 1151–1172.
Seaman, S. R. and Richardson, S. (2004). Equivalence of prospective and retrospective models in the Bayesian analysis of case-control studies. *Biometrika* **91**, 15–25.
Severini, T. L. (2001). *Likelihood Methods in Statistics.* Oxford: Oxford University Press.
Silvey, S. D. (1959). The Lagrange multiplier test. *Ann. Math. Statist.* **30**, 389–407.
Silvey, S. D. (1961). A note on maximum likelihood in the case of dependent random variables. *J. R. Statist. Soc.* B **23**. 444–452.
Silvey, S. D. (1970). *Statistical Inference.* London: Chapman and Hall.
Smith, R. L. (1989). A survey of nonregular problems. *Bull. Int. Statist. Inst.* **53**, 353–372.
Song, P. X.-K., Fax, Y. and Kalbfleisch, J. D. (2005). Maximization by parts in likelihood inference (with discussion). *J. Am. Statist. Assoc.* **100**, 1145–1167.
Stein, C. (1959). An example of wide discrepancy between fiducial and confidence intervals. *Ann. Math. Statist.* **30**, 877–880.
Stigler, S. (1990). *The History of Statistics: The Measurement of Uncertainty before 1900.* Cambridge, Mass: Harvard University Press.
Stone, M. (1969). The role of significance testing: some data with a message. *Biometrika* **56**, 485–493.
Storey, J. D. (2002). A direct approach to false discovery rates. *J. R. Statist. Soc.* B **64**, 479–498.
Sundberg, R. (1974). Maximum likelihood theory for incomplete data from an exponential family. *Scand. J. Statist.* **1**, 49–58.
Sundberg, R. (1994). Precision estimation in sample survey inference: a criterion for choice between variance estimators. *Biometrika* **81**, 157–172.
Sundberg, R. (2003). Conditional statistical inference and quantification of relevance. *J. R. Statist. Soc.* B **65**, 299–315.
Thompson, M. E. (1997). *Theory of Sample Surveys.* London: Chapman and Hall.
Thompson, S. K. (1992). *Sampling.* New York: Wiley.
Todhunter, I. (1865). *History of the Theory of Probability.* Cambridge: Cambridge University Press.
Todhunter, I. (1886, 1893). *History of the Theory of Elasticity.* Edited and completed for publication by K. Pearson. Vols 1 and 2. Cambridge: Cambridge University Press.
van der Vaart, A. (1998). *Asymptotic Statistics.* Cambridge: Cambridge University Press.
Vaagerö, M. and Sundberg, R. (1999). The distribution of the maximum likelihood estimator in up-and-down experiments for quantal dose-response data. *J. Biopharmaceutical Statist.* **9**, 499–519.
Varin, C. and Vidoni, P. (2005). A note on composite likelihood inference and model selection. *Biometrika* **92**, 519–528.
Wald, A. (1947). *Sequential Analysis.* New York: Wiley.
Wald, A. (1950). *Statistical Decision Functions.* New York: Wiley.
Walley, P. (1991). *Statistical Reasoning with Imprecise Probabilities.* London: Chapman and Hall.

Wedderburn, R. W. M. (1974). Quasi-likelihood function, generalized linear models, and the Gauss–Newton method. *Biometrika* **61**, 439–447.

Welch, B. L. (1939). On confidence limits and sufficiency, with particular reference to parameters of location. *Ann. Math. Statist.* **10**, 58–69.

White, H. (1994). *Estimation, Inference, and Specification Analysis.* New York: Cambridge University Press.

Whitehead, J. (1997). *Design and Analysis of Sequential Clinical Trials.* New York: Wiley.

Williams, D. (2001). *Weighing the Odds: A Course in Probability and Statistics.* Cambridge: Cambridge University Press.

Yates, F. (1937). *The Design and Analysis of Factorial Experiments.* Technical Communication 35. Harpenden: Imperial Bureau of Soil Science.

Young, G. A. and Smith, R. L. (2005). *Essentials of Statistical Inference.* Cambridge: Cambridge University Press.

Author index

Aitchison, J., 131, 175, 199
Akahira, M., 132, 199
Amari, S., 131, 199
Andersen, P.K., 159, 199
Anderson, T.W., 29, 132, 199
Anscombe, F.J., 94, 199
Azzalini, A., 14, 160, 199

Baddeley, A., 192, 199
Barnard, G.A., 28, 29, 63, 195, 199
Barndorff-Nielsen, O.E., 28, 94, 131, 132, 199
Barnett, V., 14, 200
Barnett, V.D., 14, 200
Bartlett, M.S., 131, 132, 159, 200
Berger, J., 94, 200
Berger, R.L., 14, 200
Bernardo, J.M., 62, 94, 196, 200
Besag, J.E., 16, 160, 200
Birnbaum, A., 62, 200
Blackwell, D., 176
Boole, G., 194
Borgan, Ø., 199
Box, G.E.P., 14, 62, 200
Brazzale, A.R., 132, 200
Breslow, N.E., 160, 200
Brockwell, P.J., 16, 200
Butler, R.W., 132, 175, 200

Carnap, R., 195
Casella, G.C., 14, 29, 132, 200, 203
Christensen, R., 93, 200
Cochran, W.G., 176, 192, 200
Copas, J., 94, 201
Cox, D.R., 14, 16, 43, 63, 94, 131, 132, 159, 160, 192, 199, 201, 204
Creasy, M.A., 44, 201

Daniels, H.E., 132, 201
Darmois, G., 28
Davies, R.B., 159, 201
Davis, R.A., 16, 200
Davison, A.C., 14, 132, 200, 201
Dawid, A.P., 131, 201
Day, N.E., 160, 200
Dempster, A.P., 132, 201
de Finetti, B., 196
de Groot, M., 196
Dunsmore, I.R., 175, 199

Edwards, A.W.F., 28, 195, 202
Efron, B., 132, 202
Eguchi, S., 94, 201

Farewell, V., 160, 202
Fisher, R.A., 27, 28, 40, 43, 44, 53, 55, 62, 63, 66, 93, 95, 132, 176, 190, 192, 194, 195, 202
Fraser, D.A.S., 63, 202
Fridette, M., 175, 203

Garthwaite, P.H., 93, 202
Gauss, C.F., 15, 194
Geisser, S., 175, 202
Gill, R.D., 199
Godambe, V.P., 176, 202
Good, I.J., 196
Green, P.J., 160, 202
Greenland, S., 94, 202

Hacking, I., 28, 202
Hald, A., 194, 202

Author index

Hall, P., 159, 203
Halmos, P.R., 28, 203
Heyde, C.C., 194, 203
Hinkley, D.V., 14, 43, 131, 132, 201, 202
Hochberg, J., 94, 203

Jaynes, E.T., 94, 203
Jeffreys, H., 44, 131, 196, 203
Jenkins, G.M., 199
Jensen, E.B.V., 192, 199
Jiang, W., 177, 203
Johnson, V.E., 131, 203

Kadane, J.B., 202
Kalbfleisch, J.D., 93, 159, 203
Kass, R.E., 94, 131, 203
Keiding, N., 199
Kempthorne, O., 192, 203
Keynes, J.M., 195
Kolmogorov, A.N., 65
Koopmans, B.O., 28

Laird, N.M., 201
Lange, K., 132, 203
Laplace, P.S. de, 194
Lawless, J.F., 175, 203
Lee, Y., 175, 203
Lehmann, E.L., 29, 44, 63, 203
Liang, K.Y., 176, 203
Lindley, D.V., 62, 93, 94, 131, 196, 203
Lindsay, B., 160, 204
Liu, J.S., 132, 204

Manly, B., 192, 204
Marchetti, G., 132
Mayo, D.G., 44, 204
McCullagh, P., 131, 160, 204
Mead, R., 132, 204
Medley, G.F., 132, 201
Meng, X-L., 132, 204
Mitchell, A.F.S., 94, 204
Mondal, D., 16, 200
Murray, M.K., 131, 204
Mykland, P.A., 160, 204

Nelder, J.A., 132, 175, 203, 204
Neyman, J., 25, 29, 43, 175, 194, 195, 204

O'Hagan, A., 202
Owen, A.B., 160, 204

Pawitan, Y., 14, 93, 204
Pearson, E.S., 25, 29, 63, 176, 194, 195, 204
Pearson, K., 63, 204
Pitman, E.J.G., 28, 176
Prentice, R.L., 160, 204
Pyke, R., 160, 204

Raftery, A.E., 131, 203
Rao, C.R., 14, 176, 204
Ramsey, F.P., 196
Reid, N., 132, 160, 192, 199–201, 204
Rice, J.A., 14, 131, 204
Richardson, S., 160, 205
Ripley, B.D., 132, 204
Ritz, C., 159, 204
Robert, C.P., 132, 204
Robinson, G.K., 93, 204
Romano, J.P., 29, 63, 203
Ross, G.J.S., 16, 204
Rotnitzky, A., 159, 204
Royall, R.M., 28, 195, 205
Rubin, D.B., 193, 201, 205

Savage, L.J., 28, 196, 203
Seaman, S.R., 160, 205
Seneta, E., 194, 203
Severini, T.L., 14, 205
Silverman, B.W., 160, 202
Silvey, S.D., 14, 131, 159, 199, 205
Skovgaard, I.M., 159
Smith, A.F.M., 62, 200
Smith, R.L., 14, 159, 205, 206
Snell, E.J., 63, 201
Solomon, P.J., 16, 201
Song, P.X.-K., 160, 205
Sprott, D.A., 159, 203
Stein, C., 94, 205
Stigler, S., 194, 205
Stone, M., 43, 205
Storey, J.D., 94, 205
Sundberg, R., 93, 94, 132, 192, 205

Takeuchi, K., 132, 199
Tamhane, A., 94, 203

Thompson, M.E., 192, 205
Thompson, S.K., 192, 205
Tiao, G.C., 14, 62, 200
Todhunter, 194
Turnbull, B., 177, 203

Utts, J., 93, 200

Vaagerö, M., 94, 205
van der Vaart, A., 131, 205
van Dyk, D., 132, 204
Varin, C., 160, 205
Vidoni, P., 160, 205

Wald, A., 94, 163, 176, 195, 205

Walley, P., 93, 205
Wang, J.Z., 159, 203
Wasserman, L., 94, 203
Wedderburn, R.W.M., 160, 206
Welch, B.L., 63, 206
Wermuth, N., 131, 132, 201
White, H., 159, 206
Whitehead, J., 94, 206
Williams, D., 14, 206
Winsten, C.B., 199

Yates, F., 40, 192, 206
Young, G.A., 14, 206

Zeger, S.L., 176, 203

Subject index

acceptance and rejection of null hypotheses, *see* Neyman–Pearson theory
adaptive quadrature, 127
adequacy of model, *see* model criticism
admissible decision rule, 163
analysis of variance, 186
ancillary statistic, 47, 48, 57
   extended, 49
   sample size as, 89
anecdotal evidence, 82
asymptotic relative efficiency of tests, 176
asymptotic theory, 100
   Bayesian version of, 106, 107
   comparison of alternative procedures in, 117
   equivalence of statistics in, 103, 105
   fictional aspect of, 100
   higher-order, 128, 132
   multidimensional parameter in, 107, 114
   nuisance parameters in, 109
   optimality of, 105
   regularity conditions for, 130
   standard, 100, 151
   sufficiency in, 105
autoregression, 12
axiomatic approach, 62, 65

Bartlett correction, 130
Bayes factor, 131
Bayes rule, 162
Bayesian inference, 191
   advantage and disadvantage of, 64, 96
   asymptotic theory of, 106
   case-control study, for, 160
   classification problem in, 91
   disagreement with frequentist solution of, 90
   empirical Bayes, 64, 75, 79, 81
   formal simplicity of, 46
   full, 81
   general discussion of, 9, 10, 45, 47, 59, 62
   implementation of, 79, 83, 96
   irregular problems for, 159
   model averaging, 84, 117
   model comparison by, 115
   modified likelihood, role in, 160
   multiple testing in, 87
   numerical issues in, 127
   prediction, treatment of, 161
   randomization in, 191, 193
   semi- or partially, 111
   sequential stopping for, 89
   test, 42–44
   unusual parameter space for, 143
best asymptotically normal (BAN) test, 177
beta distribution, 74
betting behaviour, used to elicit probability, 72
bias, assessment of, 82
BIC, 116
binary data, regression for §5, 171 *see also* binomial distribution, two-by-two contingency table
binary fission, 20, 119
binomial distribution, 5, 21, 23, 38, 51, 53, 54, 74, 85
birth process, *see* Binary fission
Bonferroni, 87
bootstrap, 128, 160
Brownian motion, limitations of as a model, 174

canonical parameter, *see* exponential family
canonical statistic, *see* exponential family
case-control study, 154–157, 160
Cauchy–Schwarz inequality, 164, 176
censoring, *see* survival data
Central Limit Theorem, 27, 130, 151, 180, 187
chi-squared, 29, 60, 108
choice-based sampling, *see* case-control study
classification problem, various treatments of, 91, 93
coherency, 72
   domination by relation to real world, 79
   temporal, 78, 79
combination of evidence, 81
completely randomized design, 185
completeness, 63, 167
complex models, approach to analysis of, 173
component of variance, 11, 16, 61, 80
   caution over use of, 62
conditional inference
   model formulation by, 53, 54
   role in interpretation of, 71, 86
   technical, 54
   unacceptable, 56
   see also ancillary statistic, sufficient statistic
confidence distribution, 66
confidence ellipsoid, 27
confidence interval, 26
   desirability of degenerate cases, 41, 143
   interpretation of, 8
   likelihood-based, 27
   significance test related to, 40, 41, 115
   unconditional and conditional contrasted, 48
confidence set, 104, 133
conjugate direction, 92, 95
conjugate prior, 74, 75
contingency table, 121
   see also two-by-two contingency table
continuous distributions, dangers of, 137
convergence in probability, 131
covariance matrix, 4, 27, 121
covariance selection model, 123
criticism, *see* model criticism

Cramér–Rao inequality, 164, 165
cumulant, 28, 141
cumulant generating function, 21
curved exponential family, *see* exponential family

data mining, 14
data quality, crucial importance of, 1, 86
data-dependent modification of analysis, 86
data-dependent prior
   dangers of, 78
   inevitability of, 78
decision analysis, 7, 72, 162
   approaches to, 162, 176
   classification problem for, 91
   role of utility in, 162
degree of belief, *see* impersonal degree of belief, personalistic probability, Bayesian inference
density
   deplorable definition of, 2
   irrelevance of nonuniqueness, 15
design of experiments, 184–192
   design-based analysis, 186, 192
   factorial arrangement, 146
   model-based analysis, 185
design-based inference, 178
directly realizable factor of likelihood, 149
discrepancy between asymptotic equivalents, 105
discreteness in testing, 25
discriminant function, 92, 95
dispassionate assessment, frail attempt at, 1–196
dispersion index, for Poisson distribution, 33
distribution-free test, 37

Edgeworth expansion, 132
efficacy of test, 166
efficiency of alternative estimates, 166
elicitation, *see* Bayesian inference
EM algorithm, 127, 132
empirical likelihood, 160
empirical logistic transform, 171
empirically weighted least squares, 170
entropy, 76
envelope method, 126

Subject index

escape from likelihood pathology, 134
estimating equation, *see* unbiased estimating equation
estimating vector, 157
estimation of variance, critique of standard account of, 168
Euler's constant, 113
exchange paradox, 67, 93
expectation, properties of, 14, 15
expert opinion, value and dangers of, 82, 93
explanatory variable, conditioning on, 1, 2
exponential distribution, 5, 19, 22
 displaced, 137
exponential family, 20, 23, 28, 96
 Bayesian inference in, 85
 canonical parameter constrained, 122
 canonical parameter in, 21
 canonical statistic in, 21
 choice of priors in, 23
 curved, 22, 23, 121
 frequentist inference for, 50, 51
 incomplete data from, 132
 information for, 98
 mean parameter, 21
 mixed parameterization of, 112
exponential regression, 75

factorial experiment, 145
failure data, *see* survival data
false discovery rate, 94
fiducial distribution, 66, 93, 94
 inconsistency of, 67
Fieller's problem, 44
finite Fourier transform, 138
finite population correction, 180
finite population, sampling of, 179, 184
Fisher and Yates scores, 40
Fisher's exact test, *see* hypergeometric distribution, two-by-two contingency table
Fisher's hyperbola, 22
Fisher information, *see* information
Fisher's identity, for modified likelihood, 147
Fisherian reduction, 24, 47, 68, 69, 89
 asymptotic theory in, 105
frequentist inference
 conditioning in, 48
 definition of, 8
 exponential family, for, 55
 formulation of, 24, 25, 50, 59, 68
 need for approximation in, 96
 relevance of, 85
 role as calibrator, 69
 rule of inductive behaviour, 70
fundamental lemma, *see* Neyman–Pearson theory

gamma distribution, 56
generalized hypergeometric distribution, 54
generalized method of moments, 173, 177
Gothenburg, rain in, 70, 93
gradient operator, 21, 28
grid search, 125
group of transformations, 6, 58

hazard function, 151
hidden periodicity, 138
higher-order asymptotics, *see* asymptotic theory
history, 194–196, *see also* Notes 1–9
hypergeometric distribution, 180, 190
 generalized, 53
hyperprior, 81

ignorance, 73
 disallowed in personalistic theory, 72
 impersonal degree of belief, 73, 77
importance sampling, 128
improper prior, 67
inefficient estimates, study of, 110
information
 expected, 97, 119
 in an experiment, 94
 observed, 102
 observed preferred to expected, 132
information matrix
 expected, 107, 165
 partitioning of, 109, 110
 singular, 139
 transformation of, 108
informative nonresponse, 140, 159
innovation, 13
interval estimate, *see* confidence interval, posterior distribution
intrinsic accuracy, 98

invariance, 6, 117
inverse gamma distribution as prior, 60, 62
inverse Gaussian distribution, 90
inverse probability, *see* Bayesian inference
irregular likelihoods, 159

Jeffreys prior, 99
Jensen's inequality, 100

Kullback–Leibler divergence, 141

Lagrange multiplier, 122, 131
Laplace expansion, 115, 127, 129, 168
Laplace transform, *see* moment generating function
Least squares, 10, 15, 44, 55, 95, 110, 157
Likelihood, 17, 24, 27
    conditional, 149, 160
    conditions for anomalous form, 134
    exceptional interpretation of, 58
    higher derivatives of, 120
    irregular, 135
    law of, 28
    local maxima of, 101
    marginal, 149, 160
    multimodal, 133, 135
    multiple maxima, 101
    partial, 150, 159
    profile, 111, 112
    sequential stopping for, 88
    unbounded, 134
    see also modified likelihood
likelihood principle, 47, 62
likelihood ratio, 91
    signed, 104
    sufficient statistic as, 91
likelihood ratio test
    nonnested problems for, 115
    see also fundamental lemma, profile log likelihood
linear covariance structure, 122, 132
linear logistic model, 171
linear model, 4, 19, 20, 55, 145, 148
linear regression, 4
linear sufficiency, 157
location parameter, 5, 48, 57, 73, 98, 129
log normal distribution, 74
logistic regression, 140

Mantel–Haenszel procedure, 54, 63
Markov dependence graph, 123
Markov Property, 12, 119, 152
Markov chain Monte Carlo (MCMC), 128, 132
martingale, 159
matched pairs, 145, 146, 185
    nonnormal, 146
maximum likelihood estimate
    asymptotic normality of, 102, 108
    definition of, 100
    exponential family, 22
    Laplace density for, 137
    properties of, 102
mean parameter, *see* exponential family
metric, 29
missing completely at random, 140
missing information, 127
mixture of distributions, 144
model
    base of inference, 178
    choice of, 114, 117
    covering, 121
    failure of, 141, 142
    nature of, 185
    primary and secondary features of, 2
    saturated, 120
    separate families of, 114
    uncertainty, 84
model criticism, 3, 7, 37, 58, 90
    Poisson model for, 33
    sufficient statistic used for, 19, 33
modified likelihood, 144, 158, 159
    directly realizable, 149
    factorization based on, 149
    marginal, 75
    need for, 144
    partial, 159
    pseudo, 152
    requirements for, 147, 148
moment generating function, 15, 21
multinomial distribution, 33, 53, 63
multiple testing, 86, 88, 94
multivariate analysis, normal theory, 6, 29, 92

Nelder–Mead algorithm, 126, 132
Newton–Raphson iteration, 126

Neyman factorization theorem, *see* sufficient statistic
Neyman–Pearson theory, 25, 29, 33, 36, 43, 63, 68, 163, 176
    asymptotic theory in, 106
    classification problem for, 92
    fundamental lemma in, 92
    optimality in, 68
    suboptimality of, 69
non central chi-squared, paradox with, 74
non-likelihood-based methods, 175
    see also nonparametric test
non-Markov model, 144
nonlinear regression, 4, 10, 16, 22, 139
nonparametric model, 2
nonparametric test, 37
normal means, 3, 11, 32, 46, 56, 59, 165
    Bayesian analysis for, 9, 60, 73, 80
    consistency of data and prior for, 85
    information for, 98
    integer parameter space for, 143
    ratio of, 40
notorious example, 63, 68
    related to regression analysis, 69
nuisance parameter, *see* parameters
null hypothesis, 30
    see significance test
numerical analysis, 125, 132

objectives of inference, 7
observed information, *see* information
odds ratio, 5
one-sided test,
optional stopping, *see* sequential stopping
orbit, 58
order of magnitude notation, 95, 131
order statistics, 20, 40
    nonparametric sufficiency of, 38
orthogonal projection, 157
orthogonality of parameters
    balanced designs, in, 112

p-value, *see* significance test
parameter space
    dimensionality of, 144
    nonstandard, 142, 144
    variation independence, 2
parameters
    criteria for, 13, 14, 112
    nuisance, 2

    of interest, 2
    orthogonality of, 112, 114
    superabbundance of, 145–147
    transformation of, 98, 99, 102, 108, 131
    vector of interest, 27
parametric model, 2
    see also nonparametric model, semiparametric model
partial likelihood, 150, 152
periodogram, 187, 189
permutation test, 38, 138
personalistic probability, 79, 81
    upper and lower limits for, 93
Pitman efficiency, *see* asymptotic relative efficiency
personalistic probability, *see* Bayesian inference
pivot, 25, 27, 29, 175
    asymptotic theory, role in, 109
    irregular problems, for, 136
    sampling theory, in, 181
plug-in estimate for prediction, 161
plug-in formula, 161, 162
point estimation, 15, 165–169
Poisson distribution, 32, 34, 55, 63, 99, 147
    multiplicative model for, 54, 63
    overdispersion in, 158
Poisson process, 90
    observed with noise, 41, 124
posterior distribution, 5, 9
power law contact, 136
prediction, 84, 161, 175
predictive distribution, 161
predictive likelihood, 175
primary feature, *see* model
prior closed under sampling, *see* conjugate prior
prior distribution, 9
    consistency with data of, 77, 85
    flat, 73
    improper, 46
    matching, 129, 130
    normal variance, for, 59
    reference, 76, 77, 83, 94
    retrospective, 88
    see also Bayesian inference
probability
    axioms of, 65
    interpretations of, 7, 65, 70

probability (*contd*)
   personalistic, 71, 72
   range of applicability of, 66
profile log likelihood, 111, 119
projection, 10
proportional hazards model, 151
protocol, *see* significance test
pseudo-score, 152
pseudo-likelihood, 152, 160
   binary sequence for, 153
   case-control study for, 154
   pitfall with, 154
   time series for, 153
pure birth process, *see* binary fission

quadratic statistics, 109
quasi likelihood, 157, 158, 160
quasi-Newton method, 126
quasi-score, 158

random effects, 146
   see also component of variance
random sampling without replacement, 55, 179, 180
random walk, 119
randomization, motivation of, 192
randomization test, 188, 189, 191
randomized block design, 185
rank tests, 39
Rao–Blackwellization, 176
ratio estimate in sampling theory, 182, 184
Rayleigh distribution, 22
recognizable subset, 71, 93
rectangular distribution, *see* uniform distribution
reference prior, *see* prior distribution
region of Neyman structure, 63
regression, 16
regulatory agency, 70
rejection of null hypotheses, *see* Neyman–Pearson theory
residual sum of squares, 10, 19, 145, 172
Riemannian geometry, 131

saddle-point, 133, 159
   expansion, 132
sample size, data-dependent choice of, 89
sampling theory, 179–184, 192

sandwich formula, 142, 153
scale and location problem, 6, 56, 58
   reference prior for, 77
scale problem, 6
score, 97
score test, 104
secondary feature, *see* model
selection effects in testing, 78
selective reporting, 86, 87
self-denial, need for, 81
semi-asymptotic, 124
semiparametric model, 2, 4, 26, 151, 160
sensitivity analysis, 66, 82, 175
separate families, 142, 159
   likelihood ratio for, 142
sequential stopping, 88, 90, 94
Shannon information, 94
Sheppard's formula, 154
sign test, 166
   asymptotic efficiency of, 167
significance test, 30
   choice of test statistic, 36, 37
   confidence intervals related to, 31
   discrete cases, 34, 43
   interpretation of, 41, 42
   linear model for, 49
   nonparametric, 40, 44
   one-sided, 35, 36
   protocol for use of, 86
   severity, 44
   simple, 31, 32
   strong null hypothesis, 186
   two-sided, 35, 36
   types of null hypothesis, 30, 31
simulation, 3, 127, 132, 159, 174, 175
singular information matrix, 139, 141
smoothing, 4
spatial process, 12, 16
standard deviation, unbiased estimation of, 167
Stein's paradox, 74
strong Law of Large Numbers, essential irrelevance of, 29
Student t distribution, 8, 26, 59, 61, 169, 190
sufficient statistic, 18, 20, 28
   ancillary part, 48
   Bayesian inference, role in, 18
   complete, 50
   factorization theorem for, 18
   minimal form of, 18

motivation for, 18
Neyman–Pearson theory, related to, 68
use of, 24
superpopulation model, 181
support of distribution, definition of, 14
survival data, 39, 113, 149, 159
survival function, 113, 151
symmetry, nonparametric test of, 38
systematic errors, 66, 94
see also bias

tangent plane, 12, 158
temporal coherency, *see* coherency
time series, 12, 138, 159
transformation model, 57, 59, 63
transformation of parameter, *see* parameters
transparency of analysis, 172
two measuring instruments, random choice of, *see* notorious example
two-by-two contingency table
  exponential family based analysis of, 51, 54
  generating models for, 51, 63, 190
  Poisson distribution, relation to, 52
  randomization test for, 190, 191
two-sample nonparametric test, 39

unbiased estimate, 15, 163, 169
  construction of, 167–169
  correlation between different estimates, 165
  exceptional justification of, 164
  exponential family, in, 165
unbiased estimating equation, 163
  modified likelihood, for, 147
uniform distribution, 20, 47, 135
unique event
  probability of, 70
up and down method, 94
utility, 162

validity of modified likelihood estimates, 148
variance ratio used for model criticism, 36
variation independence, 49

weak law of large numbers, 27, 29, 79
Weibull distribution, 113
  displaced, 137

You, definition of, 71

Printed in the United States
By Bookmasters